The Book of
INDIAN TREES

The Book of
INDIAN TREES

K.C. Sahni

BOMBAY NATURAL HISTORY SOCIETY

OXFORD
UNIVERSITY PRESS

OXFORD

UNIVERSITY PRESS

Oxford University Press is a department of the University of Oxford. It furthers the University's objective of excellence in research, scholarship, and education by publishing worldwide in

Oxford   New York

Auckland   Bangkok   Buenos Aires   Cape Town   Chennai
Dar es Salaam   Delhi   Hong Kong   Istanbul   Karachi   Kolkata
Kuala Lumpur   Madrid   Melbourne   Mexico City   Mumbai   Nairobi
São Paulo   Shanghai   Taipei   Tokyo   Toronto

Oxford is a registered trademark of Oxford University Press
in the UK and in certain other countries

First edition 1998
Second edition 2000
Reprint 2005
Reprint 2010

© K.C. Sahni 2000

ISBN 0 19 564589 8

Cover photograph: Purple Bauhinia *Bauhinia purpurea* © Isaac Kehimkar
Cover design: V. Gopi Naidu and Divya Fernandez

This book has been set in Times New Roman by J.P.K. Menon at Bombay Natural History Society, Mumbai.

PRINTED BY BRO. LEO AT ST. FRANCIS INDUSTRIAL TRAINING INSTITUTE, MOUNT POINSUR, BORIVLI (W), MUMBAI 400 103, PUBLISHED BY THE BOMBAY NATURAL HISTORY SOCIETY, HORNBILL HOUSE, SHAHEED BHAGAT SINGH ROAD, MUMBAI 400 023, AND CO-PUBLISHED BY MANZAR KHAN, OXFORD UNIVERSITY PRESS, YMCA LIBRARY BUILDING, JAI SINGH ROAD, NEW DELHI 110 001.

CONTENTS

Pages

Foreword to the Second Edition vii

Preface ix

Acknowledgements xi

List of Plates xii

Introduction 1

Description of species * 16

Know your Trees 207

Glossary 215

Bibliography 221

Index of Scientific Names 223

Index of Common Names 227

---

* All text figures have been reduced to 80% of actual size

FOREWORD TO THE SECOND EDITION

The importance of trees has assumed great significance since the UN Conference on Environment, held in Stockholm in 1972. This subject is now a global problem and was discussed at length at the Rio de Janeiro UN Conference on Environment, twenty years after Stockholm. Their great value is now being realised in view of the ecological imperatives such as the urgent need to control pollution and avert global warming.

The Book of Indian Trees by Professor K.C. Sahni, Bombay Natural History Society/ Oxford University Press, 1998, is a concise version of INDIAN TREES by Brandis. It covers seven countries of the Subcontinent, India, Pakistan, Bangladesh, Nepal, Bhutan, Myanmar and Sri Lanka. The last named is not covered by Brandis. The author has carried out floristic surveys in the area covered, mostly while he was posted at the Forest Research Institute, Dehra Dun, where he was Director, Biological Research. It comprises information that has accumulated since the publication of the work by Dr. Brandis. The book contains a detailed introduction. The major part of the book is devoted to individual species, field identification, description, features of the bark, leaves, flowers, fruits, seeds, distribution and recurring seasonal events like leaf fall, flowering and fruiting. Details on propagation and etymology are given. Trees not included in the species account that occur in the eleven different regions of the Subcontinent are highlighted.

Nomenclature of species is up to date, vernacular and English names are provided. As mentioned, the work covers indigenous trees. *Jacaranda* and others have been excluded as they are exotics and are described in Blatter and Millard's SOME BEAUTIFUL INDIAN TREES and FLOWERING TREES by M.S. Randhawa. Eucalypts are also excluded as they are exotics and are described in THE WEALTH OF INDIA. This was necessary in order to include as many native trees as possible.

Another dimension is added to the book when the author describes the relationship between certain animals and their favourite trees, such as tiger tree whose bark is used by tigers for cleaning their claws, the majestic karayani of the 'Ghats' of Kerala being the favourite haunt of the lion-tailed macaque, and the giant buttressed jungly dungy providing unique nesting holes for the great hornbills. Another is the talipot palm that develops a bud which bursts with a loud pop, releasing the largest known inflorescence in the Plant Kingdom.

The items under "Miscellaneous" pertain to uses, legends, folklore and ecological notes. The volume is further enlivened by seventeen coloured plates by renowned photographers, and seventy line drawings and eleven silhouettes by Mr. P. N. Sharma, one of the leading botanical artists of our country. The "Know your Trees" is user friendly, giving field and diagnostic characters which aid spot recognition.

Lucidly written, the book complements Dr. Sálim Ali's acclaimed treatise on Indian birds.

The author is a distinguished botanist and was FAO/United Nations Forest Botanist, when he wrote IMPORTANT TREES OF THE NORTHERN SUDAN. He was a recipient of the *Environment Day Award* in 1994, followed by the *Seth Memorial Award* by the Indian Society of Tree Scientists in 1995. The work is well-timed, as the Government has spoken of the need to make environment awareness a compulsory subject in schools and colleges.

We should thank the Department of Science and Technology and the Purshotamdas Thakurdas and Divaliba Charitable Trust, Mumbai, for providing financial support to keep the cost reasonable. The Bombay Natural History Society and the Oxford University Press, deserve our gratitude for bringing out this attractive book.

I am confident that this 2nd edition fills a real need for a comprehensive and authoritative book, as evidenced by the fact that the first edition in 1998 was sold out in a year's time. This second edition is an update of the first with changes in nomenclature, additions in descriptions and a colour photograph of *Magnolia campbellii* from the Eastern Himalaya, one of the world's most spectacular trees.

M.S. Swaminathan, FRS, FNA

M.S. Swaminathan Research Foundation
Chennai
December 30, 1999

PREFACE

THE BOOK OF INDIAN TREES is intended as a popular guide describing the characteristics and distribution of trees of the Indian subcontinent. It embraces a vast region of S. Asia and SE. Asia covering seven countries: India, Pakistan, Bangladesh, Nepal, Bhutan, Myanmar and Sri Lanka. The choice of trees is confined to important, indigenous, conspicuous and interesting trees of natural history interest such as karayani (*Cullenia exarillata*) of the 'ghats' of Kerala which is the habitat of the lion-tailed macaque (*Macaca silenus*), upas tree (*Antiaris toxicaria*), the poisonous milky latex of which is used for poisoning arrows for hunting, tiger tree (*Bischofia javanica*) the soft bark of which is favoured by tigers for cleaning their claws, its trunk often deeply scored with claw marks, and many such trees associated with animals and birds; also included are spectacular trees such as *Magnolia campbellii* starring the mountain slopes of the Eastern Himalaya, bonfire tree, dotting the hillsides of peninsular India, Sri Lanka and Myanmar, like masses of flaming red coral, rhododendrons which dominate the landscape of the Himalaya, Nilgiris and the highlands of Sri Lanka, hollock (*Terminalia myriocarpa*), tree ferns, palms, canes, bamboos, conifers, cycads, etc.

Foreign trees such as jacaranda, gulmohur, etc. have been excluded as these are described in Blatter & Millard's **Some Beautiful Indian Trees** published by the Bombay Natural History Society, also **Flowering Trees** by Randhawa. Eucalypts have been excluded as they are exotics and are described in the '*Wealth of India*' published by the Council of Scientific and Industrial Research, New Delhi. Their exclusion was necessary in order to include as many indigenous trees as possible in this concise work. The chief purpose of this work is to provide a means for the layman to become acquainted with the main trees of our forests and thus foster an interest in forestry and natural history.

In the Introduction a brief account of the forest vegetation, geography, climate and hints on identification are given. The information on species includes the latest valid botanical name, with well known synonyms, popular name, vernacular names in use in the countries covered; etymology indicating how the botanical name is derived; a systematic and comprehensive botanical description; phenological data giving the months of leafing, flowering and fruiting to aid seed collection, details of distribution of trees, 'spot characters' by which is meant the outstanding diagnostic characters of species for quick identification, uses and propagation. This book is profusely illustrated to aid recognition of trees described in this work.

Every tree has a botanical name and a popular name. Botanical names are always written in Latin. It is the *lingua franca* of botanists of the world. Latin names are used and recognized throughout the world and under the International Code of Botanical Nomenclature (ICBN) the same name cannot be correctly applied to more

i x

than one species. If two or more botanical names are applied to the same tree, the oldest correct name has priority and is valid.

The layman rarely uses the botanical name. Instead he uses the popular name. One of the purposes of this book is to popularise and supply standard meaningful common or popular names which can be used throughout the area covered by this work and outside, such as the W. Himalayan low level fir for *Abies pindrow* which is not found in the E. Himalaya; strawberry tree for *Benthamidia capitata,* the fruits of which are strawberry-like; S. Indian soapnut or S. Indian ritha for *Sapindus emarginatus* which is found in S. India.

The interest in wildlife, both flora and fauna, is becoming more and more evident among our educated classes. In this activity the Bombay Natural History Society has played an outstanding role in popularising and conserving wildlife under the leadership of Dr. Sálim Ali. During the last half century the mantle of trees that covered much of the subcontinent has gradually reduced, owing to a rising population and the consequent demand for fuel, fodder and timber for building. In order that the environment does not deteriorate any further, the Ministry of Environment and Forests has been very active to preserve core areas of our forests abounding in flora and fauna. To date 15 Biosphere Reserves, 157 National Parks and 659 Sanctuaries have been demarcated in the different biogeographical zones of the country. This has also been done in the countries covered in this work.

This book is full of original observations taken during my tours in the subcontinent including Sri Lanka. Forest floras are available for most states of India and districts, e.g. The Forest Flora of Bombay Presidency and Sind by W.A. Talbot, I.F.S., Forest Flora for Chakrata, Dehra Dun and Saharanpur Forest Divisions by U.N. Kanjilal revised by B.L. Gupta, **Indian Trees** by D. Brandis, I.F.S., and many others. The last named is the standard work of reference to trees, shrubs and woody climbers of the whole Subcontinent excluding Sri Lanka. Interested readers who wish to advance their knowledge further for specific states or areas may refer to such works.

As I have taught Forest Botany to Indian Forest Service probationers at the Indian Forest College, Dehra Dun, for over 25 years, this book should be most useful to all levels of forestry students, viz., I.F.S., State Forest Service Officers, Rangers' Colleges as well as forestry students in Pakistan, Bangladesh, Bhutan, Nepal, Myanmar and Sri Lanka. This work should be of special interest to students at the Wildlife Institute, as adequate examples of associated animals and birds are cited.

K. C. Sahni

ACKNOWLEDGEMENTS

My thanks are due to the Bombay Natural History Society for inviting me to write this book. Acknowledgement is made to Mr. P. N. Sharma for undertaking the exacting task of botanical illustrations and I hope that readers will find his drawings useful in identification. These drawings are mostly based on collections in the herbarium, Botanical Garden and Arboretum of the Forest Research Institute, Dehra Dun. Though photocredits have been mentioned, I would also like to thank the photographers.

K.C. Sahni

The Bombay Natural History Society acknowledges with gratitude financial support for the publication of this book from the Department of Science and Technology, Government of India, and the Purshotamdas Thakurdas & Divaliba Charitable Trust, Mumbai.

LIST OF PLATES

Plate No. **Species** (number denotes Serial No. in description)

1.	2.	Campbell's Magnolia *Magnolia campbellii*
	4.	White Champa *Michelia doltsopa*
2.	21.	Red Silk Cotton *Bombax ceiba*
3.	25.	Karaya *Sterculia urens*
4.	26.	Bonfire Tree *Firmiana colorata*
5.	28.	Rudraksh *Elaeocarpus augustifolius*
6.	51.	Indian Laburnum *Cassia fistula*
7.	54.	Flame of the Forest *Butea monosperma*
8.	68.	Hollock *Terminalia myriocarpa*
9.	88.	Strychnine Tree *Strychnos nux-vomica*
10.	99.	Tiger Tree *Bischofia javanica*
11.	114.	Silver Birch *Betula utilis*
12.	130.	Tiger Bamboo *Bambusa vulgaris*
13.	138.	Chir Pine *Pinus roxburghii*
14.	139.	Chilgoza Pine *Pinus gerardiana*
15.	145.	East Himalayan Fir *Abies densa*
16.	147.	West Himalayan Spruce *Picea smithiana*
17.		**Silhouettes of Palms**
	121.	Date Palm *Phoenix dactylifera*
	124.	Palmyra palm *Borassus flabellifer*
	125.	Fish Tail Palm *Caryota urens*
	127.	Coconut Palm *Cocos nucifera*
18.		**Silhouettes of Evergreens**
	144.	Fir *Abies pindrow*
	149.	Himalayan Larch *Larix griffithiana*
		Drooping Juniper *Juniperus recurva*
		Black Juniper *Juniperus indica*
19.		**Silhouettes of deciduous trees**
	39.	Horse Chestnut *Aesculus assamica*
	15.	Gurjan *Dipterocarpus turbinatus*
	21.	Red Silk Cotton *Bombax ceiba*
20.	65.	*Rhizophora mucronata* with stilt roots
	66.	*Bruguiera gymnorhiza* buttressed trees

INTRODUCTION

This book covers the countries of the Indian subcontinent India, Pakistan, Bangladesh, Nepal and Bhutan. Myanmar and Sri Lanka are also touched, albeit briefly.

Included are over 153 representative native trees covering broad-leaved trees, palms, screw pines, bamboos, Gymnosperms (cycads, yews, conifers, etc.) and tree ferns. Those not covered in the detailed species account, but are fairly prominent in the different regions, are very briefly described in this introductory chapter.

The choice of trees has been limited to common, wild and interesting trees to rouse interest in natural history. The subcontinent has approximately 2000 kinds of trees. Foreign or exotic trees like jacaranda, gulmohur, Java cassia, etc., which have been introduced have been excluded. These are described and illustrated in colour in E. Blatter and W.S. Millard's well-known book **Some Beautiful Indian Trees** (revised by W.T. Stearn, 1977). Eucalypts have been excluded as they are exotics and described in forestry books and in *The Wealth of India* (CSIR, New Delhi). Their exclusion was necessary in order to include as many indigenous trees as possible.

The account is in popular language, as the book is meant for the general intelligent reader, not the botanist. It has not been possible to avoid the use of botanical terms entirely, as such an attempt often makes the phraseology awkward and clumsy. A glossary has been provided to overcome difficulties in understanding botanical terms. With the aid of colour transparencies, line drawings (which are provided for each tree), description and spot characters, the identification of trees should become easy. Popular English names have been chosen or coined which help in recognition, e.g. tiger bamboo or golden bamboo for *Bambusa vulgaris* var. *vittata*, fish tail palm for *Caryota urens*, 'bidi' tree for *Diospyros melanoxylon*, the leaves of which are used for making Indian cigarettes or 'bidis', etc.

As the recognition of trees consists largely in knowing what to look for, a few hints may be given. Trees are divisible into the following major heads: (1) Broad leaved (*Sal, Kadam*, etc.), (2) Monocotyledons (palms, screw pines, Bamboos), (3) Gymnosperms meaning naked seeds not enclosed in an ovary (conifers, i.e. cone bearers, viz. pines, *Deodar*, etc.; cycads, yew and cypress), (4) Tree Ferns, which are a prominent feature of the landscape of the Eastern Himalaya, Nilgiris and the highlands of Sri Lanka.

1

Broad-leaved trees form the major component of our vegetation and are divided into a large number of families which are described in this book. The botanist identifies a tree by its flowers. The first step in identification is to see if the petals are free or united or if there are no petals. If they are free as in a rose, they fall in the first group of families called POLYPETALAE, No. 1 Dilleniaceae to No. 31 Cornaceae. If the petals are fused or united as in the Rhododendrons, they fall in the second group of families called GAMOPETALAE, i.e. No. 32 Clusiaceae to No. 40 Verbenaceae. The third group APETALAE, when petals are absent or when the sepals and petals are similar in form and texture as in the Camphor flower, i.e. families No. 41 Lauraceae (Camphor family) to No. 52 Salicaceae (Willow and Poplar family).

Characters of the leaves are also helpful in identification. They may be arranged in pairs on either side of the stem, and leaves so arranged are said to be opposite, e.g. teak. The commonest arrangement is for the leaves to be alternate, e.g. the *Pipal* tree. The leaf may consist of one continuous expanse or it may be cut up into separate pieces. In the former it is said to be simple, e.g. mango, in the latter compound, e.g. *Amaltas*. In compound leaves the leaflets may be arranged in two rows on the common axis, e.g. tamarind. Such leaves are said to be pinnate or resembling a feather. If in a leaf of the pinnate type we get in place of leaflets secondary axes which bear leaflets pinnately arranged, the leaf is said to be twice pinnate or bipinnate, e.g. *Babul*.

If in a compound leaf the leaflets are borne on the common stalk at one point, the leaf is said to be palmate or resembling the fingers of the hand, e.g. *Semal* or red silk cotton tree. A common type of compound leaf has 3 leaflets, and is called trifoliate, e.g. the *Bael* tree. Leaves may show glands seen as translucent dots when held against the light and are best seen under a hand lens. These are seen in members of the Lemon family, the *Bael* tree or *Jamun*. Leaves and twigs may contain milky juice as in the *Pipal* tree. This is seen by plucking the leaf stalk or by a cut on the stem. If yellow milky latex oozes out that is typical of the family Clusiaceae e.g. *Poeciloneuron indicum*. Vernacular names are helpful in identification. If unable to recognize a tree in a forest, one can ask the local people for its vernacular name and then look up the corresponding botanical name, its description and spot characters in the local flora to establish its identity. Failing that one may contact the relevant circle of the Botanical Survey of India or the Forest Botanist, Forest Research Institute, Dehra Dun, or send them pressed specimens with flowers, fruits and foliage by post to know the name of the tree.

Botanists confirm the identity of plants by comparison with authentically identified and preserved specimens in the Herbarium. The Herbarium of the Forest Research Institute at Dehra Dun is air-conditioned to ensure that specimens are well preserved for posterity. The collections are so exhaustive that it is difficult to find an Indian plant that is not there. This herbarium is one of the best in Asia and specializes in trees and other arboreal flora. The other is the Central National Herbarium of the Botanical Survey of India at Calcutta. This is the oldest herbarium in India, having all categories of plants including lower plants such as mosses and liverworts. Revision of the Flora of India, which is one of the richest in the world is underway at this prestigious herbarium and at the Regional Herbaria of the Botanical Survey of India.

The interested field naturalist should carry a good pocket lens, a jack-knife or 'khukri', binoculars, polythene bags for collecting specimens and a forest flora and field note book. If the trees are very tall he should look for fallen flowers, fruit and foliage on the forest floor. With some practice he will be able to recognise many trees by means of their field characteristics. Cuts with a 'khukri' or knife are prohibited in arboreta to avoid damage to trees but a mild blazing of the trunk is allowed in the forests. Some of the field characteristics that help identification are covered in a separate chapter **Know your Trees** towards the end of this concise volume.

It is suggested that this work is used in conjunction with *Indian Trees* by D. Brandis (1921) which covers the countries covered in the present work except Sri Lanka. To these may be added works of wide coverage, such as *Encyclopedia of Horticulture* by L.H. Bailey (1950) and *The Oxford Encyclopedia of Trees of the World* by Bayard Hora (1981) to foster interest in trees.

Forest Vegetation: The plant life of the subcontinent is unique in the context of world flora for its richness. It is the second richest in the world after Brazil. The richness of the flora is due to the vastness of the area, embracing so many degrees of latitudes and its range of climates and topography. There is every type of climate and habitat from deserts (hot deserts of Sind and Rajasthan and cold deserts of Ladakh at 3,650-5,180 m) and the dry scrub of the Punjab to tropical evergreen rain forests of Assam, Western Ghats, Andaman and Nicobar Islands, and the coniferous and broad-leaved forests of the Himalaya. It is convenient to recognise 11 different regions for the purposes of this book.

1.   WESTERN HIMALAYA

This region stretches from West Nepal through Kumaon to Kashmir, to the Murree Hills and beyond in Pakistan.

Typical trees are listed altitudewise :

(a)   **Submontane up to 1,500 m**: Sal *Shorea robusta*, flame of the forest *Butea monosperma*, red silk cotton tree *Bombax ceiba*, shisham *Dalbergia sissoo*, jamun *Syzygium cumini*, safed siris *Albizia procera*, toon *Toona ciliata*, amla *Emblica* *officinalis*, haldina *Adina cordifolia*, sandan *Ougeinia oojeinensis*, Indian laburnum *Cassia fistula*, and semla *Bauhinia semla* syn. *B. retusa*, (Semla is after its Hindi name).

(b)   **Subtropical evergreen to 1,500 m**: More characteristic of the drier Pakistan Himalaya, Kashmir and Himachal Pradesh. Wild Pomegranate *Punica granatum*, Olive *Olea ferruginea*.

(c)   **Temperate 1,500 to 3,400-3,700 m**: Lower reaches bear chir pine *Pinus roxburghii*, grey oak *Quercus leucotrichophora*, alder *Alnus nepalensis*, etc. Higher up: Blue pine *Pinus wallichiana*, deodar *Cedrus deodara*, *Rhododendron arboreum* (absent in Kashmir valley), horse chestnut *Aesculus indica*, Himalayan poplar *Populus ciliata*, Himalayan elm *Ulmus wallichiana*, Himalayan cypress *Cupressus torulosa* in Garhwal and Kumaon on limestone rocks and maple *Acer caesium*. Western Himalayan fir *Abies pindrow* forms forests of large extent between 2300-3300 m, yew *Taxus baccata* is scattered in occurrence. The edible chilgoza pine *Pinus gerardiana* occurs in the inner dry valleys of Himachal Pradesh, Chitral and North Baluchistan in areas of high winter snowfall and scanty rainfall. Himalayan pencil juniper 'dhup' *Juniperus polycarpos* is also found in the inner semi-arid valleys in Lahaul, Kagan Valley and in Baluchistan forming forests of considerable extent. The upper reaches are formed by the brown oak *Quercus semecarpifolia* (absent in Kashmir Valley), the high level fir *Abies spectabilis* occurs between 3,300-3,700 m. The silver birch or bhojpatra tree *Betula utilis* forms the upper limit of tree vegetation all along the Himalaya.

The West Himalayan ranges differ from the Eastern in greater length, higher latitude, cooler-drier climate and in the far greater breadth of the mountain mass. A transverse section drawn through the valley of Kashmir, from the plains of the Punjab to the Karakorum is 3 times as long as one drawn anywhere transversely across the Eastern Himalaya (Joseph Hooker, **A Sketch of the Flora of British India**, Clarendon Press, Oxford, 18-19, 1906). The extreme NW. Himalaya in Kashmir and Pakistan is devoid of the nuisance of leeches which pester the tourists in the Eastern Himalaya!

2.   **EASTERN HIMALAYA**

Central Nepal through North Bengal, Sikkim, Bhutan to Arunachal Pradesh. The Eastern Himalaya is far more evenly humid than the Western Himalaya because of the proximity of the former to the Bay of Bengal, the world's largest bay. In the Darjeeling, Sikkim, Bhutan, Arunachal corner, the Himalaya is horseshoe shaped owing to the bending of the mountains and thus it catches the bulk of the monsoon-bearing clouds. The monsoon is so intense here that the Siwaliks which run from the Punjab to Arunachal Pradesh have almost entirely been eroded

4

away over the ages, for 322 km, from the Kosi to Manas in Bhutan. Sikkim is a Himalayan name which means "the land of lightning" because of continual flashes of lightning in the skies. Mist-enshrouded for half of the year and blessed with heavy rainfall, the Eastern Himalaya provides an ideal home for spectacular rhododendrons and exquisitely beautiful orchids. High humidity is conducive to tree growth and consequently the timber line or upper limit of trees in the East Himalaya is up to 4,570 m as compared to 3,600 m in the West Himalaya.

The Eastern Himalayan foothills have a contrasting vegetation to the Western. It is characterised by nutmegs Myristicaceae, screw pine *Pandanus*, giant buttressed trees of jungli dungy *Tetrameles nudiflora* a favourite nesting tree of hornbills, the handsome hollock tree *Terminalia myriocarpa* with masses of tiny cream flowers turning into coppery red, small winged fruits in autumn and the tiger tree *Bischofia javanica*, whose bark is favoured by tigers for cleaning their claws. Tree ferns, palms and bamboo are conspicuous. There are 26 species of bamboos which go on decreasing westwards and totally disappear in the Kashmir valley proper. The West Himalayan panorama is dominated by vast and gregarious coniferous forests of chir, blue pine, deodar and fir. In the East Himalaya the coniferous forests are not vast but scattered, though exceeding in number of genera and species. The typical East Himalayan conifers are Himalayan larch *Larix griffithiana*, hemlock *Tsuga dumosa*, East Himalayan fir *Abies densa*, East Himalayan spruce *Picea spinulosa*, plum-yew *Cephalotaxus griffithii*, *Podocarpus neriifolius*, etc. Moisture loving oaks, laurels, white champa *Michelia doltsopa*, Rhododendrons and Magnolia are common, Campbell's magnolia *M. campbellii* starring the mountain slopes with magnificent snow white flowers in spring when it is leafless. Rhododendrons under giant hemlocks and firs dominate the landscape of the higher reaches.

a)   **Tropical zone up to 1,500 m**: Typical trees are sal *Shorea robusta*, red silk cotton *Bombax ceiba*, lampatia *Duabunga grandiflora* with nocturnal white flowers emitting a smell of sour milk, kadam *Anthocephalus sinensis*, elephant apple *Dillenia indica*, gamari *Gmelina arborea*, tiger tree *Bischofia javanica*, jungli dungy *Tetrameles nudiflora*, hollock *Terminalia myriocarpa*, *Acrocarpus fraxinifolius* (green petals, sepals and crimson stamens), golden champa *Michelia champaca*, kala dammar *Canarium strictum*, ironwood *Mesua ferrea*, white champa *Michelia doltsopa*, heart flower *Talauma hodgsonii*, petsut *Engelhardia spicata*, makrisal *Schima wallichi*, alder *Alnus nepalensis*, sangli-kung *Betula cylindrostachys* (handsome tree, bark peeling off in flakes), rhinoceros bamboo *Dendrocalamus hamiltonii*, etc.

b)   **Temperate 1,800-3,500 m**: Tree ferns *Cyathea gigantea*, *C. brunoniana*, buk oak *Quercus lamellosa*, safed champa *Michelia doltsopa*, pipli *Exbucklandia* *populnea,* Campbell's magnolia *M. campbelli, Rhododendron arboreum, R. falconeri* (flowers cream coloured, with purple spots at the base within), etc.

Above 2,700 m: Blue Pine (absent in North Bengal and Sikkim), Himalayan hemlock, Himalayan larch and Eastern Himalayan Spruce.

Above 3,000 m: Himalayan larch, East Himalayan fir, black juniper *Juniperus wallichiana* and bhojpatra.

c) **Subalpine 3,500-4,500 m.**: The most dominant tree is East Himalayan Fir or Red Fir *Abies densa*. The other associates black juniper *Juniperus wallichiana* and Himalayan silver birch are comparatively scarce, the latter forming the upper limit of tree vegetation. *On the upper slopes*: Many species of shrubby Rhododendrons with several shades of blood red, creamy-white, lilac, cinnamon-red are a most spectacular sight with a backdrop of snow covered mountain ranges.

In the East Himalaya, there are 82 species of rhododendrons, mostly shrubs, as compared to only five in the West Himalaya out of which only *R. arboreum* is a tree.

3.   NORTH WEST DRY REGION

Sind, Baluchistan, West Punjab, plains of North West Frontier Province, East Punjab, Haryana, Rajasthan and Gujarat.

Khejri *Prosopis cineraria*, pilu *Salvadora oleoides*, tamarisk *Tamarix aphylla* (feathery foliage, pink flowers), babul *Acacia nilotica* subsp. *indica*, black siris *Albizia lebbeck*, shisham, dhaukra *Anogeissus pendula*, salai *Boswellia serrata*, gum karaya tree *Sterculia urens*, hingu *Balanites aegyptiaca* (fruits when stuffed with gunpowder are used as crackers during the Diwali festival), gum Arabic tree *Acacia senegal*, ber or Indian jujube *Ziziphus* spp., Rajasthan teak *Tecomella undulata* (trumpet shaped orange-red flowers), Indus poplar *Populus euphratica* (grows along the river Indus), etc. Two palms, the wild date *Phoenix sylvestris* and the mazari palm *Nannorhops ritchieana* are typical of Sind, the Salt Range and the last named around the Khyber Pass. Mangroves in the Indus delta repeat the vegetation of the Sunderbans but with a reduced number of species, mostly *Rhizophora* and *Sonneratia*.

4.   GANGETIC PLAIN

Plains of West Uttar Pradesh to Hooghly delta and adjacent littoral forests of Sunderbans. Khejri, babul, kunlai *Dichrostachys cinerea* and sal; flanking the Aravalis, *Anogeissus pendula*, salai, etc. In Bengal, the coconut *Cocos nucifera*, golden champa *Michelia champaca*, red silk cotton tree, etc. *Littoral forests of*

6

Sunderbans: Tulip Tree *Thespesia populnea* (tulip-like yellow flowers with a purple centre and black turban-shaped capsules), coral tree *Erythrina variegata* (Coral-like crimson flowers). *Mangrove*: *Heritiera minor* (Sundri, after which Sunderbans are named), *Excoecaria agallocha* (blinding tree, feared by wood-cutters: the white latex from the cut branches causes blindness), dtilt rooted mangrove *Rhizophora mucronata, Bruguiera gymnorrhiza* (the buttressed and largest of the mangroves), *Sonneratia apetala* with hanging branches and white flowers and the stemless palm, *Nypa fruticans* with feathery leaves up to 9 m long.

5. WEST COAST

South Gujarat to Kanyakumari. A region of heavy rainfall, excessively humid and mountains running parallel to the west-coast from South Gujarat to Kanyakumari with the low country between the mountains and the coast. The mountainous area is called the Sahyadris (Western Ghats). It covers South Gujarat, Maharashtra, Karnataka, Kerala and Tamil Nadu and is divisible into 2 zones of altitude; (a) the tropical zone up to *c*. 1,500 m, and (b) the temperate zone above 1,500 m. The tropical may be divided into 3 main types of forests: (1) Rain forests or wet evergreen, (2) Semi-evergreen, and (3) Moist-deciduous.

a. **Tropical zone up to *c*. 1,500 m.**

(i) **Rain Forests**: Characterized by a most luxuriant vegetation. The trees are in several storeys, the highest often buttressed. The undergrowth is often a tangle of canes and palms. Typical trees: ennai *Dipterocarpus indicus* (a magnificent Dipterocarp reaching 35 m), *Hopea* spp. with crimson winged fruits, Spar Tree *Calophyllum polyanthum* (used for masts of sailing ships), karayani *Cullenia exarillata* (food tree of the rare lion-tailed monkey), panchonta *Palaquium elipticum* (a lofty tree with a fluted stem, blaze exudes copious milk, used as gutta-percha), Indian copal tree *Vateria indica* (a handsome tree with bright red young leaves and fragrant white flowers), black dammar *Canarium strictum* (resin used as incense), jungli dungy *Tetrameles nudiflora* (interesting for its disjunct distribution, also in NE. India), *Mesua ferrea* (ironwood, wood heavy and hard, flowers like white roses, young leaves scarlet), *Pterygota alata* (very lofty buttressed tree with big round fruits 12 cm across), jack fruit *Artocarpus heterophyllus*, havalgi *Acrocarpus fraxinifolius*, many nutmegs Myristicaceae, laurels Lauraceae, etc. In Malabar, the talipot palm *Corypha umbraculifera* is most outstanding. It develops a gigantic bud over a metre high. The bud bursts with a loud pop and releases a majestic inflorescence 6 m tall by 9-12 m across. The inflorescence containing 60 million flowers is the largest known amongst flowering plants, and after flowering the palm dies.

(ii) **Semi-evergreen**: Red silk cotton, havalgi *Acrocarpus fraxinifolius*, Indian Copal and some more rain forest trees. There are several trees of moist deciduous forests, such as true laurel *Terminalia crenulata* (beautifully figured wood), kinjel *Terminalia paniculata*, nana *Lagerstroemia lanceolata* (yellowish-white bark, showy flowers in large pyramidal clusters, also called Benteak), rosewood *Dalbergia latifolia*, haldina, etc. *Terminalia bellerica*, bijai sal *Pterocarpus marsupium* (fruit like the pouch of the Kangaroo, bark exudes blood-red juice on cutting), irul *Xylia dolabriformis* (a lofty and valuable timber tree), etc. The chief bamboos are the giant thorny bamboo *Bambusa bambos* (*Bambusa arundinacea*) and the male bamboo *Dendrocalamus strictus*, solid and known for its strength and much favoured by the Indian Police!

Apart from the above there are other types of a local character representing edaphic and seral types.

(i) **Mangrove**: Dudhi baen *Avicennia alba, Sonneratia apetala*, blinding tree *Excoecaria agallocha*, etc.

(ii) **Littoral**: Alexandrian laurel *Calophyllum inophyllum*, cashewnut *Anacardium occidentale*.

(iii) **Freshwater swamps**: Nutmeg *Myristica* sp., karalli *Carallia brachiata* (the name of the tree is derived from the Telugu name, bark corky and thick, wood with handsome silver-grain), screw pine, etc. A transition zone connecting the tropical and temperate is exemplified by the hill resort of Mahabaleshwar which has two prominent trees, ajnani *Memecylon edule* (with brilliant blue flowers, a colour otherwise unknown in native Indian trees) and the myrobalan *Terminalia chebula* (yields a valuable tanning material).

(b) **Temperate Forests above 1,500 m** formed by the Nilgiris, Anaimalai and Palni Hills. The Nilgiris form the apex of the Western Ghats where they attain their greatest elevation at Anaimudi 2,600 m. The peaks are flattened or gently sloping and can be likened to an elephant's head. They rise precipitously from the west to vast grassy downs and table lands seamed with densely wooded gorges locally known as *sholas*. They contain thickly wooded evergreen forests of short-boled trees with dense rounded crowns.

Typical trees: Nilgiri rhododendron *Rhododendron arboreum* subsp. *nilagiricum*, Nilgiri champa *Michelia nilagirica*, pala *Xantolis tomentosa* (a spiny tree with yellow berries and white flowers), holly *Ilex* sp. (a tree typical of Europe and the Himalaya, the red fruit and prickly foliage are in demand during Christmas), kaymone *Ternstroemia japonica* (pale yellow flowers and fruits with red seeds). The interesting feature of the Nilgiri flora is its affinity with the flora of the Khasi, Naga Hills and Manipur. Kaymone and *Gaultheria fragrantissima* are common to both. An oil similar to oil of wintergreen is distilled from the leaves of the latter. The tree ferns *Cyathea gigantea* and *C. brunoniana* are also common to both. The ground flora is formed of *Nilgirianthus*, an undershrub with blue flowers, which come into gregarious flowering at well set intervals of 3 to 12 years or more, like an alarm clock set to go off at a certain time. The Blue Mountains, as the Nilgiris are called, are named after this plant.

The many similarities in the flora and fauna of the Western Ghats and North-east India have attracted the attention of both botanists and zoologists. This observation was later to develop into a hypothesis called the Satpura Hypothesis put forward by the noted ichthyologist Sunder Lal Hora (*Proc. Nat. Inst. Sc.* 158: 389-422, 1949) visualizing a high level connection between NE. and SW. India through the great hill range, the Satpura Hills of Central India.

6.   CENTRAL INDIAN REGION

Teak, bahera, haldina, bijai sal, rosewood, satinwood *Chloroxylon swietenia* (wood valued for ornamental work), mahua *Madhuca longifolia* var. *latifolia*, bonfire tree *Firmiana colorata* (dotting the hillsides like masses of flaming red coral), red silk cotton, salai, gum karaya, yellow silk cotton tree or buttercup tree *Cochlospermum religiosum* (brilliant golden flowers, gum used for thickening ice cream, branches used as torches in villages: being resinous they burn brightly), Flame of the Forest *Butea monosperma* (so named because the massed crown of brilliant orange flowers suggest a forest in flames). The male bamboo *Dendrocalamus strictus* is the common bamboo. Sal is widespread in Central India. Sal forests present a striking sight when covered with masses of pale yellow fragrant blooms when leafless.

a)   **Tropical thorn forest**: Babul, khair *Acacia catechu* (heart-wood dark-red which yields *Katha* used with pan or betel leaf and is the origin of splashes of dark-red which decorate our bazaars), khejri and hingu in the driest tracts.

b)   **Mangroves** on the delta of the Godavari. Stilt rooted mangrove *Rhizophora mucronata*, kankra *Bruguiera gymnorhiza*, the blinding tree, etc. Within this region fall the two great hill ranges of Central India — the Vindhya and the Satpura, and the plateau of Pachmarhi is located in the latter.

7.   DECCAN & CARNATIC

Peninsular India south of the Godavari, in Maharashtra, Andhra Pradesh, Karnataka and Tamil Nadu. Divisible into (a) an elevated hilly plateau called the Deccan sub-region, and (b) the lowland along the east (Coromandel) coast, the Carnatic sub-region.

a)   **Deccan Sub-region**: Khejri, umbrella acacia *Acacia planifrons*, axle-wood *Anogeissus latifolia*, satinwood, red sanders *Pterocarpus santalinus* (endemic to Cuddapah in Andhra Pradesh, wavy grained wood used for musical instruments and exported to Japan), sandalwood *Santalum album* (strongly scented heartwood, once a royal tree of Mysore), rosewood (gold-brown to rose purple or deep purple heartwood, streaked with black, finest wood for furniture), black siris *Albizia lebbeck, Shorea talura* (after the Tamil name 'talura', flowers fragrant ivory-like, disjunct distribution also in Myanmar, Indo-China and Thailand), *Shorea tumbuggia* (after the Tamil name 'tambugai'), etc.

b)   **Carnatic sub-region**: Palla *Manilkara hexandra* (fruit sweet and tasty, wood used in panelling, maulsiri *Mimusops elengi* (named after its Malayalam name 'elengi', fruit edible, oil used in perfumery, true ebony *Diospyros ebenum* (not enough in the Deccan, frequent in Sri Lanka, heartwood jet black used in carving), nux-vomica tree *Strychnos nux-vomica* (fruits round, hard, of the size and colour of an orange, seeds source of the drug strychnine), anjani, etc.

8.   ASSAM & MEGHALAYA

Characterized by heavy rainfall, Cherrapunji with 11,400 mm per annum holds the world record. Dense evergreen rain forests (wet evergreen) and semi-evergreen forests characterise the landscape. Leaves mostly with a pointed tip called the 'drip tip', an adaptation to quickly drain off rain water. Pointed leaves are characteristic of plants from high rainfall areas.

a)   **Rainforests (wet evergreen)**: Hollong *Dipterocarpus macrocarpus* (a towering tree to 46 m, glaringly white to grey trunk, fruit like a shuttle-cock with 2 wings), makai *Shorea assamica* (a towering giant to 50 m and 7 m in girth, flowers cream coloured, fruits shuttlecock-like with 9 cm long wings), Ironwood with white rose-like flowers, jutli *Altingia excelsa* (tall aromatic tree), chaplash *Michelia* sp., and kala dammar *Canarium strictum* (with fragrant resin). There are several bamboos, the commonest being rhinoceros bamboo *Dendrocalamus hamiltonii* (so called because the rhizome is an exact replica of the rhino horn). Hollong occurs on the south bank of the Brahmaputra because by the time the fruits drift to the north bank, they lose their viability. Short viability is characteristic of the Dipterocarpaceae or sal family.

b)   **Tropical semi-evergreen**: Cinnamomum (camphor), kala dammar, jamun, heart flower *Talauma hodgsonii* (flowers heart-shaped, aromatic, new leaves red, old leaves huge to 40 cm long, wood used for 'khukri' handles), queen's flower *Lagerstroemia reginae* (flowers like crepe paper), elephant apple *Dillenia indica* (flowers white 20 cm across, fruits large, relished by elephants), letkok *Pterygota alata*, gamari *Gmelina arborea*, Assam rubber tree *Ficus elastica*, hollock

Terminalia myriocarpa, etc. In the riverine areas: kadam, red silk cotton, and phul hingri *Sloanea assamica* (fruit 5 cm long, studded with long spines).

c)   **Hill Forests**: They approximate to those of the Eastern Himalaya except that there is no alpine zone. *Magnolia*, champa *Michelia*, maples, *Pyrus pashia* (*Sohshur* in Khasi), *Rhododendron arboreum*, Pipli *Exbucklandia populnea*, Alder *Alnus nepalensis*, low level birch *Betula alnoides* and many oaks. The pine forests occur at 760 m and above, commonly at 1,200-1,400 m and consist of only one pine, the khasi pine *Pinus kesiya*. The trees and shrubs are identical with many in the Nilgiris or are closely related.

9.   ANDAMAN & NICOBAR ISLANDS

i.   **Andamans**: A chain of 204 islands. The highest peak in North Andamans is the Saddle Peak 70 m. Narcondam Island is cone-like, the peak is 86 m high. Tall, white, pillar-like trunks of *Dipterocarpus* or gurjan stand out conspicuously as one approaches these islands by ship.

a)   **Mangroves**: *Rhizophora mucronata* standing on stilt roots, the buttressed and large *Bruguiera gymnorrhiza* followed by the gregarious and stemless palm *Nypa fruticans* with very long feathery leaves, and by a narrow belt of the local betel nut *Areca triandra*.

b)   **Littoral**: Bullet Wood *Manilkara littoralis*, a fine large tree, it forms a protective belt against the force of the southwest monsoon, wood red, hard and durable, not attacked by white ants. The Alexandrian Laurel *Calophyllum inophyllum* with its handsome foliage with numerous parallel nerves, white bombway *Terminalia procera*, Indian coral tree *Erythrina variegata*, tulip tree *Thespesia populnea*, screw pine *Pandanus odoratissimus* (flowers fragrant, fruits 15-25 c. long, scarlet), horse tail tree or whistling pine *Casuarina equisetifolia* (so called because the foliage is like a horse tail, and the fruits produce a whistling sound in the breeze).

c)   **Evergreen Forest**: Top storey, gurjan or long-leaf gurjan *Dipterocarpus grandiflorus* (very large, fruit largest in the genus with 15-25 cm long wings), chaplash *Calophyllum soulattri* (lal chini or the Nicobar canoe tree), lal bombway *Planchonia andamanica*, foliage dense, turns red before falling and hence the popular name. The local name *Bombway* is derived from the Burmese, 'Bombwe' for *Careya arborea* which resembles this tree but does not occur here. Flowers 5 cm long, white tinged with pink; madaw *Garcinia xanthochymus*, handsome tree with angled branchlets, bark dark coloured, cut yellowish, leaves up to 40 cm, the yellow acid fruits are edible.

11

d)   **Semi-evergreen**: Gurjan *Dipterocarpus alatus*, the loftiest tree up to 55 m, chief top storey tree, flowers yellowish, fruit with papery wings, Buddha's coconut tree or letkok *Pterygota alata*, tall buttressed tree to 45 m, fruits of the size and colour of a coconut, papita *Pterocymbium tinctorium* with bunches of red fruits when leafless, white chuglam *Terminalia bialata* to 47 m with thin curved buttresses, butterfly shaped fruits with 2 wings, koko or black siris *Albizia lebbeck*, excellent furniture wood, padauk *Pterocarpus dalbergioides*, exuding blood-red juice when cut, with large burrs on trunk which when sliced with power saws yield beautifully figured veneers, and pyinma *Lagerstroemia hypoleuca*, handsome lilac flowers, leaves turn reddish when falling, wood a durable canoe wood.

e)   **Deciduous forests**: Occupy small strips along the coast and shed their leaves in the hot season. White bombway, white chuglam or safed chuglam, dhup *Canarium strictum*, wood smells of vinegar, resin used by Andamanese for covering the binding of arrows, papita South Indian red silk cotton tree or dida *Bombax insigne* similar to *B. ceiba* and differs in having 400 stamens and larger petals. Jungly dungy which grows to 45 m with enormous buttresses larger than those of any in the island. The wood is used by the Andamanese for dug-out canoes, also in north-east India and Andamans marble wood *Diospyros marmorata*, wood figured with jet-black stripes, one of the most decorative timbers of the world, endemic in the Andamans. The drinking cane *Calamus andamanicus* climbing over tall trees is a source of drinking water. Sections 3 m long are cut and when held in a vertical position, the sap trickles down for some time. When the flow stops the lower 30 cm is cut and the trickle commences again.

ii)   **Nicobars**: These are 22 islands. The largest, the Great Nicobar, is 480 km south of Port Blair and 192 km north of Sumatra. It abounds in springs, streams and has five rivers. The highest peak is Mt. Thuillier (622 m). The vegetation is predominantly of the Andamans type and is characterized by the absence of gurjan *Dipterocarpus*, the principal Andamans timber and padauk. Tree ferns *Cyathea* spp. not known to occur in Andamans are common in the moist valleys of the Great Nicobar. Presence of South Indian *Podocarpus wallichianus* was recorded for the first time in 1952 from the interior of Gt. Nicobar on hillsides above Alexandra River [Sahni, K.C. Botanical Exploration in the Gt. Nicobar Island. *Indian For.* 79(1): 3-16, 1953].

Sri Lanka has close affinity with the flora of the Western Peninsula and Malabar. Its mountain flora is similar to the Nilgiri flora. The hot and dry northern districts are of the coromandel type. The umbrella-shaped *Acacia planifrons* is characteristic of northern Sri Lanka and is also seen in the hot and dry districts of Tamil Nadu. The species common to the Western Peninsula are many, e.g. the true ebony *Diospyros ebenum* which occurs in quantity, the true cinnamon *Cinnamomum verum* is the true '*dalchini*' obtained from the bark in the form of rolled quills, talipot palm *Corypha umbraculifera* has the largest inflorescence in the world, its huge leaves 4.66 m across are used as tents by Sri Lankan soldiers, fish-tailed palm *Caryota urens* and the toddy palm *Borassus flabellifer* with fan-shaped leaves with orange-coloured drupes in clusters. Toddy is drawn from the groves of toddy trees. According to connoisseurs toddy tastes like mild champagne, others associate its taste with cider. The tree rhododendron *R. arboreum* subsp. *zeylanicum* is characterised by strongly concave leaves with a blistered or puckered upper surface, lower surface with a spongy fawn indumentum (Chamberlain Notes RBG Edinb. 39 (2) : 239, 1982). The tree fern *Cyathea gigantea* is common to the mountains of Sri Lanka and the Nilgiris. However, there is a high degree of endemic flora, one such plant is a pale green endemic bamboo *Ochlandra stridula* which covers hundreds of square kilometres. It produces a creaking sound caused while treading on its stem.

Two Malaysian trees long in cultivation here are worthy of note viz. mangosteen and durian. The former is one of the most prized fruits of the tropics. Its delicious white pulp melts in the mouth like ice cream and in flavour is something between a grape and a peach. The durian is shaped like a spiked football. Its custard-like pulp mixed with sugared cream is delicious but has a putrid smell. A taste for durian is quickly acquired.

Myanmar is very rich botanically and has many trees in common with northeast India. The coastal and mangrove forests are similar to Bangladesh and West Bengal. It is divisible into 4 sub-regions. Northern Myanmar is mountainous with a few peaks above 2,941 m. The vegetation throughout is like the Eastern Himalaya. There is an additional gymnosperm, the Chinese coffin tree *Taiwania cryptomerioides*. The khasi pine *Pinus kesiya* is also present. Spruce, fir hemlock, larch and juniper are present (see Eastern Himalaya) which in the account by Sir Joseph Hooker (1906) *A Sketch of the Flora of British India*, p.47 were reported as absent as the area was under-explored at that time. Further south is found the two needle pine *Pinus merkusii* at 150-750 m, the only pine in the world that crosses the equator. The trees at higher elevations in North Myanmar besides conifers are *Rhododendron arboreum, Magnolia, Michelia, Pyrus*, oaks, etc.

Western Myanmar: Dipterocarps, bamboos, canes and oaks are prominent. Eastern Myanmar: The only Dipterocarp is *Pentacme siamensis*, the timber is heavy and durable and is known as 'Engyin' locally. A giant rose *Rosa gigantea* though not a tree is interesting because of its immense giant white flowers 12.5 cm across. The thick stems are made into walking sticks. Also in Manipur at 1,800 m. Central Myanmar: Characterized by 2 species of teak *Tectona grandis* and *Tectona hamiltoniana*. The last named is also a useful timber and occurs in the dry zone. The forests of Myanmar constituted the finest teak forests in the world at one time. These were conserved and saved from destruction by Sir Dietrich Brandis, India's first Inspector General of Forests. Brandis faced fierce opposition from the contractors' lobby, who argued that the supply of teak in Myanmar was inexhaustible. Brandis in those early times in the mid-nineteenth century laid emphasis on conservation, carried out linear valuation surveys with an estimate of the growing stock and thus saved the teak forests of Myanmar.

The most outstanding tree of Myanmar is the flame amherstia *Amherstia nobilis*, one of the most beautiful trees in the world. The 20-26 crimson flowers are borne on 1 m long candelabra-like drooping sprays. The individual flowers 20 x 10 cm across look like humming birds. Another fascinating tree is the famous upas tree *Antiaris toxicaria* found in the Western Ghats. It is a huge buttressed tree towering to 75 m with buttresses of 10 m or more up the trunk. Jungle dwellers tip arrows into the poisonous milky juice of the tree to hunt birds and animals.

Colour: The bright colouring of the young shoots of evergreen trees which delights the eye are *Mesua ferrea, Acer oblongum, Amherstia nobilis, Mangifera indica, Quercus leucotrichophora, Memecylon*, etc. The young leaves of the species of this last named genus may be pink, purple or deep-blue. The autumn colouring of deciduous species, which is a marked feature in the temperate and subtropical zones which is not uncommon in the Himalaya, are *Parthenocissus semicordata* (woody climber), chinar (*Platanus orientalis*), *Acer campbellii*, etc., is comparatively rare in the rest of India. Yet there are noteworthy exceptions such as badam (*Terminalia catappa*), *Anogeissus latifolia, A. pendula*, etc.

Associated birds and animals: These are recorded for those trees for which authentic information was available. The most outstanding examples are: The jungli dungy *Tetrameles nudiflora*, an enormous tree of the Western Ghats and Eastern Himalayan foothills is a favourite nesting tree of hornbills. *Bischofia javanica* of the sub-Himalayan tracts is favoured by tigers. Owing to its soft juicy cortex, tigers clean their claws by drawing them along the trunk. The bark is often deeply scored with their claw marks with blood red stains from the red juice of the cut trunk. In the 'ghats' of Kerala, karayani *Cullenia exarillata* forms the top storey and is the habitat of the rare lion-tailed macaque. They break open the spiny capsular fruits to eat the pulp. Destruction of the habitat has made this animal an endangered species.

The red panda *Ailurus fulgens*, a near relation of the giant panda of China, is 'one of the most beautiful animals in existence' with a chestnut coloured coat. It is found in the East Himalayan broad-leaved forest type (East Nepal to Arunachal) above 1,525 m which consists of oaks, maple, birch, magnolia, rhododendrons, fir and bamboos mainly *Arundinaria racemosa* and *Cephalastachyum capitatum*. The animal is wholly vegetarian and particularly relishes bamboo shoots. These two bamboos are edible. The giant panda, the most spectacular endangered animal of the world, was exterminated from Upper Myanmar about 40-50 years ago (Prater 1948, **The Book of Indian Animals**), owing to the destruction of its bamboo habitat. The red panda is also threatened owing to the practice of shifting cultivation, habitat destruction and opening up of Arunachal Pradesh and Sikkim under development plans. I photographed the Red Panda around Dirang much below Sela in Arunachal Pradesh. It is easily tamed and was in transit to the Itanagar Zoo when I saw it.

The wild trees described in this book may best be viewed or photographed in colour in the several National Parks, Sanctuaries and proposed Biosphere Reserves in the subcontinent. Wildlife enthusiasts should write to the Bombay Natural History Society to seek their guidance to choose areas of their preference to view flora, fauna and scenery. They may also be seen while traversing the countries of the subcontinent by rail or road. In cities like New Delhi, Mumbai, Bangalore, etc., many exotic trees are flourishing. These may be recognised with the help of M.S. Randhawa's illustrated book 'Flowering Trees'.

1. LARGE-FLOWERED DILLENIA, ELEPHANT APPLE *Dillenia indica* Linn.

(Family: Dilleniaceae)

Hindi, Bengali *Chalta*; Marathi *Karmal*; Malayalam *Chilta*; Burmese *Thabyu*.

Field Identification: Leaves in bunches at the ends of branches 15-35 x 5-13 cm, oblong-lance-shaped, pointed, toothed, with parallel lateral nerves. Flowers white, fragrant, 12-20 cm in diameter, solitary. Sepals 5, fleshy and concave. Petals 5, large and white. Stamens numerous, forming a yellow crown around the white spreading rays of the stigma. Fruits large, 7.5 to 10 cm, green. **Description:** An erect evergreen tree up to 24 m in height and up to 1.8 m in girth with a dense rounded crown. Trunk like a sculptured female figure. **Bark** reddish-brown, peeling off in papery flakes. **Leaves** in bunches at the ends of branches, 15-35 cm long and 5-13 cm broad, oblong-lance-shaped, leaf stalks 2.5 cm, leaves toothed, pointed at the apex, characterized by parallel lateral nerves which gives them a beautiful fluted surface. **Flowers** large, 12-20 cm in diameter, white, fragrant and solitary. Flower stalks club-shaped, concave, thick and fleshy. Petals 5 large, obovate. The numerous stamens form a yellow crown round the white spreading rays of the stigma. **Fruit** 7.5-12.5 cm in diameter, always green, hard, consisting of 5 closely-fitting sepals enclosing numerous kidney-shaped hairy margined **seeds** embedded in the pulp. **Distribution:** In sub-Himalayan tracts from Nepal to Arunachal Pradesh, Bangladesh, Myanmar and Sri Lanka. In Peninsular India in Maharashtra, Mysore, Kerala, Andhra Pradesh, Orissa, Bihar and Madhya Pradesh, on banks of streams. The fruits are relished and dispersed by wild elephants (hence the popular name Elephant Apple), more often by water and are carried away to germinate on banks of streams. The fallen fruits quickly decay, white ants eat the pulp, filling up the shell with earth where the seeds germinate. **Phenology:** Normally evergreen, almost leafless in Dehra Dun briefly in June. New leaves appear in July. **Flowers** July to August. **Fruit** ripens in October and continue to fall to the ground during the winter and summer. **Miscellaneous:** The thickened sepals eaten raw have the flavour of sour unripe apples; they are more agreeable when stewed with sugar and have the flavour of apples stewed in the same manner. They are often mixed in curries in Bengal, especially prawn curries, to which they impart an agreeable flavour. The juice of the fruit sweetened with sugar forms a cooling drink. The fruits are relished by elephants. The veneer of the timber is attractively figured and is used in panelling. The tassar and Atlas moth silkworms are said to feed on its leaves. The leaves are used for polishing ivory. **Gardening:** A popular ornamental tree for parks, propagated from seeds sown during the monsoon. **Etymology:** The genus was named by Linnaeus to commemorate Johann Jacob Dillenius (1684-1747), a German botanist and Professor of Botany at Oxford; '*Dillenia* has of all plants the showiest flower; Dillenius is likewise conspicuous among botanists' Linnaeus, *Critica Bot.*, 80: 1737.

2. CAMPBELL'S MAGNOLIA
Magnolia campbellii Hook. f. & Thoms.
(Family: Magnoliaceae)
Plate 1

Hindi *Lal Chanp* (pink flowers); Nepali *Ghoge Chanp* (white flowers); Bhutan *Pendder, Patagiri*.

Field Identification: Deciduous tree, 18-24 m high. Leaves 10-30 cm long with 12-16 pairs of lateral nerves. Flowers axillary or terminal, large and showy, goblet shaped, 15-25 cm across, fragrant, white or pink. Fruiting spike cylindric 15-20 x 2.5 cm. Seeds red. **Description:** A most handsome deciduous tree 18-24 m high with a wide crown 15 m across. **Bark** very dark coloured, that of the branches almost black. Blaze white. Wood greyish white, soft. **Leaves** alternate, 10-30 cm long, lateral nerves 12-16 pairs. **Flowers** axillary or terminal, large, showy and white, 15-25 cm across, goblet-shaped. Sepals and petals similar, falling off after a week, arranged in whorls of three, 12-15 in number. Flowers occasionally rose-coloured. **Fruiting** spike cylindric 15-20 cm x 2.5 cm **Seeds** red. This is one of the loveliest of Asian Magnolias of the mist enshrouded Eastern Himalaya, spectacular when leafless in bloom with nearly 5000 flowers. The forest floor covered with fallen petals adds to the spectacle. The white flowering form called *Ghoge Chanp* is found in Tonglu and Senchal in Darjeeling and in Kameng in Arunachal Pradesh. The pink flowering form called *Lal Chanp* is found in Gumpahar and in Senchal, both in Darjeeling. The flowers generally are clear pink on the outside and a suffused combination of white and lighter pink on the inside. **Distribution:** Eastern Himalaya from Nepal to Arunachal Pradesh and Manipur at 2100-3000 m. **Phenology:** Flowers in April when the tree is leafless. New **leaves** appear in May and are shed in the autumn. **Fruits** October-November. **Miscellaneous:** Planted in tea gardens for shading tea bushes. The yellowish-white wood presents a beautiful colour in panelling. Also used in making tea boxes. **Gardening:** Propagation by layering. An allied species *M. pterocarpa* according to J. Hutchinson of Kew is perhaps the most ancient species of flowering plants. This is now on the endangered list of flora and is recorded from the lower hill forests in Darjeeling and Subansiri in Arunachal Pradesh.

3. CHAMPAK, YELLOW or GOLDEN CHAMP
Michelia champaca Linn.

(Family: Magnoliaceae)

Bengali, Hindi *Champa*; Tamil, Malayalam *Champakam*; Marathi *Son champa*; Burmese *Sagawa*.

Field Identification: Tall tree with a tapering crown. Leaves wavy margined. 13-25 x 5-9 cm, lance-shaped. Flowers fragrant; 3-4.6 cm long by 5-6.4 cm across, star-shaped, hidden in the dense foliage. Sepals 15-21, oblong lance-shaped. Petals 15-21, deep yellow, the outer oblong, acute, the inner linear. Fruits a cluster of capsules on 7.5-15 cm long spike, capsule dark brown. Seeds 1-12 scarlet or brown, ultimately hanging on long cords. **Description:** A tall, handsome, evergreen tree, with a close, tapering crown up to 33 m or more, with a girth of 2.4-3.7 m. **Bark** thick, ashy-grey, dull brown inside, the blaze turning reddish-brown inside. **Leaves** wavy margined, 13-25 x 5-9 cm, lance-shaped, sometimes ovate, pointed, petiole 1.8-3 cm. **Flowers** axillary, rarely terminal, solitary, 3-4.6 x 5-6.4 cm in diameter, pale yellow to deep yellow, fragrant, double narcissus-like. Sepals 15-21, oblong or oblong lance-shaped. Petals 15-21, deep yellow or orange, the outer oblong, acute, the inner linear. **Fruits** a long cluster of capsules on a spike 7.5-15 cm long, capsules dark brown, opening by two valves. **Seeds** 1-12, polished, scarlet or brown, ultimately hanging on long cords. **Distribution:** Eastern sub-Himalaya from Nepal to Arunachal Pradesh, Naga Hills, Myanmar and the Western Ghats at 500-1,500 m in evergreen forests. **Phenology:** New **leaves** appear about March. Flowers April-May. **Fruit** ripens in August. **Miscellaneous:** The bark is browsed by deer. Parakeets and other birds devour the seeds. Bees find the perfume too heavy and pass it by. Ants, rats and squirrels destroy nursery-sown seeds, which have to be coated with red lead to prevent damage. The flowers yield champa oil, used in perfumery. The heartwood is strong and durable, capable of taking a high polish, and is valued as a furniture wood. Golden champa is widely cultivated in gardens and temples. Flowers are used in religious ceremonies and are worn in the hair by women. **Gardening:** By seeds which are sown in August soon after ripening, as they lose their viability early. A fast growing tree, attaining 2.4 m in 2 years on deep moist soil.

4. CHAMP
Michelia doltsopa Buch.-Ham in DC.

(Family: Magnoliaceae)

Syn. *M. excelsa* Bl. ex Wall.

Plate 1

Hindi, Nepali *Safed Champ*; Trade *Champ*.

Field Identification: Tall trees with ovate-elliptic or oblong lance-shaped leaves, pointed at the tip with a wedge-shaped or rounded base. Lateral nerves 8-16 pairs. Petioles swollen at the base. Flowers white, sweet scented with 12 sepals and petals outer 7.5 x 3.3 cm. Fruiting spike lax, 10-20 cm. Carpels 1.2 cm, beaked. Seeds red. **Description:** A tall tree 30-36 m in height with a girth of 3 m; up to 6.6 m girth is recorded from a tree in Singalila Range in Darjeeling. **Bark** dark grey, corky. **Leaves** 13-20 x 5-9 cm, ovate-elliptic or oblong-lance-shaped, pointed, wedge shaped or slightly rounded at the base, thinly leathery, main lateral nerves 8-16 pairs, petiole 1.5-3 cm long, swollen at base. **Flowers** white, axillary, 8-10 cm in diameter, sweet scented, sepals and petals 12, outer 7.5 x 3.3 cm, gradually narrowing towards the centre. **Fruiting** spike lax, 10-20 cm long. Carpels 1.2 cm, shortly beaked. Seeds red. **Distribution:** In the Eastern Himalaya from East Nepal to Western Arunachal Pradesh, also in Khasi and Naga Hills in Upper Hill Forests from 1500-2400 m, mostly in association with buk oak (*Quercus lamellosa*), maples and laurels. **Phenology:** Flowers mostly March-April. **Fruits** October-November. **Miscellaneous:** The red seeds are distributed by birds. Squirrels gnaw the bark and eat tips of young shoots. A first class timber in demand in Darjeeling, Arunachal Pradesh and elsewhere for furniture, doors, window frames, ceiling boards, veneers and plywood. **Gardening:** An ornamental tree worthy of introduction in parks in hill stations with high rainfall. Seeds are collected in November and sown in December as they lose viability early. They are coated with red lead to prevent attacks by rodents. The seeds germinate freely and young seedlings are transplanted a year after sowing.

5. HEART FLOWER
Talauma hodgsonii Hk. f. & Thoms.

(Family: Magnoliaceae)

Nepali *Bhalukhat*; Lepcha *Siffu-Kung*.

Field Identification: Tree with dense foliage, leaves large, leathery, deep green and shining, 20-40 x 8-15 cm with 15-25 pairs of lateral nerves, new leaves red in colour. Flowers heart-shaped, terminal, 5 x 4 cm, sepals greenish purple, petals

19

greenish white at base, bright red above, falling after opening. Fruit cone shaped 10-15 x 6-9 cm, consisting of numerous overlapping woody carpels. Follicles woody, beaked. Seeds 9 x 7 mm, red. **Description:** An evergreen tree 10-15 m tall, with dense foliage and large, deep green, shining leaves 20-40 x 8-15 cm, lateral nerves 15-25 pairs, new leaves red in colour. Petioles 2.5 cm long, thick at base. **Bark** dark brown, rough. **Flowers** terminal, 5 x 4 cm broad, sepals greenish purple, petals greenish white at base, bright red above, falling off after opening. Flowers often drop unexpanded from the tree and diffuse an aromatic smell. **Fruit** ovoid 10-15 x 6-9 cm, consisting of numerous overlapping woody carpels. Follicles woody, sharply beaked. **Seeds** oblong, plano-convex, 9 x 7 mm, bright red. **Distribution:** From E. Nepal to Arunachal Pradesh and Upper Myanmar; also in Meghalaya ascending to 1,500 m but commonly at 1,000 m. **Phenology:** The new leaves emerge 4-5 times a year and are more spectacular than the blossoms, being rich red in colour with translucence which lights up in the sunlight like a torch against the deep green background of the mature foliage. **Flowers** April-May. **Fruits** October-November. **Miscellaneous:** It is used for 'Khukri' handles. The wood is white, but black in very old trees, especially the wood of the roots, hence the Nepalese name Bhalukhat (bearwood). Sir Joseph Hooker, the great naturalist of the mid-nineteenth century, noted it as the second most beautiful flowering tree in the world. Morgan Evans, an American landscapist, introduced it in Disneyland, U.S.A. He wrote, "it is an evergreen, well clothed in huge firm leaves, producing in spring large terminal flowers comparable to the deciduous Magnolias". It deserves to be better known and is worthy of cultivation in greenhouses in the Nilgiris, Kumaon, Dehra Dun and in the hills of Sri Lanka.

6. MAST TREE
Polyalthia longifolia (Sonnerat) Thw.

(Family: Annonaceae)

Telugu, Bengali, Hindi *Devdaru*; Malayalam *Arana*; Tamil *Asoothi*.

Fig. 1

Field Identification: Evergreen tree with a pyramidal crown. Leaves 8.8-23 x 2-3.8 cm lance-shaped, tapering and with a wavy margin. Flowers hidden in the dense foliage, yellowish green, 2.5-3 cm in fascicles or short umbels. Sepals 0.5 cm long, ovate triangular. Petals 2.5 x 0.5 cm. Fruit in a cluster of 10-20, yellowish, black or purple berries 1.8-2.3 x 3-1.5 cm. Seed one, shining. **Description:** A tall evergreen with a clear, straight bole and a pyramidal crown, reaching a height of

Fig. 1

S.No. 6. *Polyalthia longifolia*. Leaves, flowers and fruits x 1.

15-18 m. **Bark** dark grey brown and smooth. **Leaves** 8.8-23 x 2-3.8 cm lance-shaped, tapering, shining, faintly gland dotted, with a wavy margin, aromatic, base wedge-shaped, petiole 0.5-1.3 cm. **Flowers** hidden in the dense foliage, yellowish green, 2.5-3.0 cm long in fascicles or short umbels. **Sepals** 0.5 cm long, ovate-triangular. **Petals** 2.5 x 0.5 cm **Fruit** in a cluster of 10-20, yellowish, black or purple berries 1.8-2.3 x 1.3-1.5 cm each containing a single seed. **Seed** smooth and shining. The var. *pendula* has a columnar crown with its branches and foliage adhering to the main stem. The tree looks like a flag-staff. **Distribution:** Indigenous to South India and Sri Lanka. **Phenology:** The new flush of leaves is completed in April. **Flowers** February-March. **Fruit** ripens July-August. **Miscellaneous:** Bats feed on ripening fruits at night. The ground is strewn with seeds the next morning, the remains of the night's banquet. The wood is used for drums in South India. It is a popular avenue tree because of its graceful appearance and dense shade. In Hindi it is sometimes wrongly called Ashok or Ashoka. The true Ashoka is *Saraca asoca*. The Mast tree is also held sacred and is planted near temples.

7. YELLOW SILK COTTON TREE, BUTTERCUP TREE, TORCHWOOD TREE
Cochlospermum religiosum (Linn.) Alston

(Family: Cochlospermaceae)

Hindi *Kumbi*; Bengali *Golgol*; Marathi *Ganglai*; Gujarati *Pahad-vel*; Tamil *Kongilam*.

Field Identification: Small tree with brilliant golden yellow flowers, 7.5-13 cm across in sub-corymbose panicles. Petals 5, up to 5 cm long, notched. Leaves palmately 3-7, commonly 5-lobed. Fruit 5-celled, leathery, pendulous, pear-shaped capsule 5-10 cm long. Seeds kidney shaped, 0.8 cm long, covered with silky floss. Leafless when in bloom. **Description:** A small deciduous tree about 6 m tall, conspicuous either when leafless with large brilliant golden yellow flowers or in full foliage with glossy-green, digitately lobed leaves, or when the large capsules are opening to disgorge the cotton-covered seeds. **Bark** fluted, 2.5 cm thick with diagonal furrows. Blaze deep brown, streaked brown and white, exuding orange-coloured sap. **Leaves** alternate, 7.5-20 cm in diameter, collected towards the ends of branches, palmately 3-7, commonly 5-lobed, lobes softly hairy, petiole 5-23 cm long, stipules linear. **Flowers** large golden yellow, 7.5-13 cm across, spectacular, short, in sub-corymbose panicles. Sepals 5, silky outside. Petals 5, up to 5 cm long, notched, broadly obovate. **Fruit** a 5-celled, furrowed, leathery, pendulous, pear-shaped capsule 5-10 cm long. **Seeds** numerous, up to 0.8 cm long, brown, kidney-shaped, pitted, covered with pale brown silky floss. **Distribution:** Western sub-Himalayan tract up to 900 m, plains of Uttar Pradesh, Bihar, West Bengal, Orissa, Madhya Pradesh, Maharashtra, Gujarat and in the Deccan east of the Western Ghats. Cultivated near temples in Sri Lanka and India. **Phenology:** The foliage is shed from November to April, new leaves appear in May-June. **Flowers** February-April. **Fruiting:** The capsules ripen and dehisce from March-July, dispersing the seeds far and wide due to the strong winds at this time of the year. The flowers last for a day and a half, they appear at 3 a.m., are wide open at sunrise and close the following morning between sunrise and noon. **Miscellaneous:** It is highly resistant to drought and forest fires. The gum from the bark is used in book binding, for cosmetics and for thickening ice cream. Because of the gum, the branches throw a clear bright flame and are used as torches in villages. The floss from the capsules is used for stuffing pillows and life belts. **Etymology:** The literal translation for the genus *Cochlospermum* means Shell Seed.

8. THE SACRED BARNA or GARLIC PEAR
Crateva magna (Lour.) DC.

(Family: Capparaceae)

Syn. *C. nurvala* Buch.-Ham., *C. religiosa* auct. non Forst. f.

Hindi *Barna*; Bengali *Barun*; Tamil *Maralingam*; Marathi *Vayavarna*.

Fig. 2

Field Identification: Small to medium-sized tree. Leaves trifoliate, leaflets lance-shaped, 8.75-17.5 x 1.5-6.5 cm, tapering. Inflorescence in lax clusters. Petals 1.5-3 x 1.5-2.2 cm with a 1.2 cm long claw. Stamens thread-like, longer than petals, purple or white when young, lilac when old. Ovary swollen, borne at the end of 5 cm long thread-like lilac-coloured stalk. Fruit woody, ellipsoid, rarely globose, 4.7-6.2 x 4-5.2 cm. Seeds dark brown, 0.5 cm across. **Description:** A small to medium-sized deciduous tree, up to 15 m in height, handsome when in full bloom. **Leaves** trifoliate and clustered at the end of the branches. Leaflets on a long stalk 4-8 cm long, lance-shaped 8.75-17.5 cm x 1.5-6.5 cm, tapering, with 7-22 pairs of lateral nerves. Inflorescence terminal, in lax clusters of about 100 flowers. Sepals green, fading to pink or yellow, 2-3.5 x 1.2-1.5 cm. Petals white fading to yellow, 1.5-3 x 1.5-2.2 cm with a claw about 1.2 cm long. A bunch of long thread-like stamens protrudes from the flower, much longer than the petals, purple or white when young, lilac when old. Barna belongs to a class of plants called gynophorous plants, which means that it bears its ovary at the end of

Fig. 2

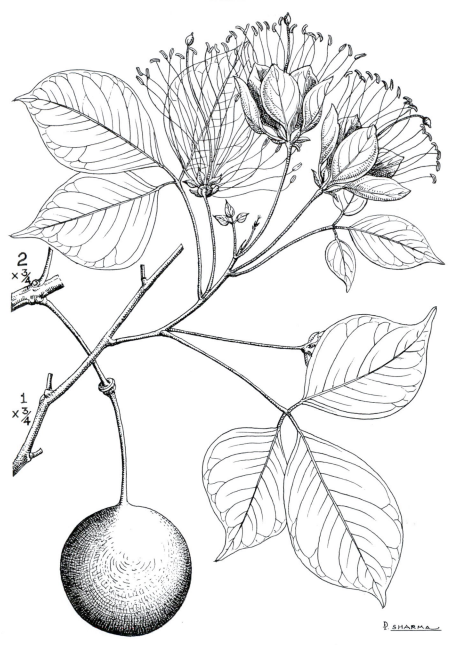

2
× ¾

1
× ¾

P. SHARMA

S.No. 8. *Crateva magna*. Leaves, flowers and fruits x ¾.

a long slender stalk, lilac in colour, thread-like, 5 cm long with the swollen ovary at the tip. When the petals fall the thread-like gynophore remains. It thickens and bears an ellipsoid (rarely globose) woody fruit 4.7-6.2 x 4-5.2 cm with numerous dark brown seeds, 0.5 cm across. **Distribution:** Deccan Peninsula (Western and Eastern Ghats), Uttar Pradesh (Gangetic plain), Bihar, Bengal, Assam, Myanmar and Sri Lanka on undulating ground or on low hills up to 600 m in shady situations near streams. **Phenology:** Leafless from November-January. New flush of leaves from February-March. **Flowering** March-May. **Fruit** ripens from June-August. **Miscellaneous:** Planted near temples and tombs as it is sacred. Wood used for making drums, toys and in planking. A cement is made by mixing the pulp of the fruit with mortar. **Gardening:** Seed is sown in the rain on deep loose soil, and transplanted during the following rains. The seed may not germinate until May or June of the year after sowing, in which case the plants will be ready for transplanting about August. **Nomenclature:** It has been erroneously named *C. religiosa* Forst in many Indian Floras, but that name belongs to a totally different Polynesian tree. **Etymology:** *Crateva* commemorates Cratevas, a Greek botanist and artist of the first century B.C.

9. KOKAM or BUTTER TREE
Garcinia indica Choisy

(Family: Clusiaceae)

Hindi *Kokum*; Marathi *Ratamba;* Kannada *Murgal*.

Field Identification: A slender tree with drooping branches. Leaves red when young, 5-10 cm long, lance-shaped or oblanceolate, nearly sessile with 6-10 pairs of lateral nerves. Fruit globose, 3.75 cm in diameter, purple, resembling a plum. Seeds embedded in red acid pulp. **Description:** A small, slender and graceful tree usually buttressed, with drooping branches, branchlets black. **Leaves** red when young, 5-10 cm long, thickly membranous lance-shaped, occasionally oblanceolate, lateral nerves 6-10 pairs, petiole 0.5-1 cm. **Flowers** small, male in terminal, 3-7 flowered, often pedunculate cymes, pedicels 0.5 cm long. Stamens numerous, anthers on short filaments crowded on a central hemispherical receptacle. Hermaphrodite flowers solitary, stamens 10-18 in 4 bundles alternating with petals. Ovary 5-7 celled, stigma 6-7 radiate, each ray with 2 lines of tubercles. **Fruit** globose, 3.75 cm in diameter, not grooved, purple, resembling a plum. **Seeds** embedded in red acid pulp, 5-8, large. **Distribution:** Endemic to the tropical rain forests of the Western Ghats, Malabar, the Konkan and Goa, often planted. **Phenology: Flowers** November-February. **Fruits** April-May. **Miscellaneous:** The seeds yield a valuable edible fat, known as kokam butter, which is medicinal. The acid in the fruit is used in cosmetics, textile, soap and other industries. The peel of the fruit is used in cooking. There is considerable trade in Goa in kokam butter. **Propagation:** It can be propagated from seed. Seeds sown at Dehra Dun in May germinated after a month. **Etymology:** The genus *Garcinia* is named after Laurence Garcin who lived, collected and wrote in India in the eighteenth century.

10. MANGOSTEEN
Garcinia mangostana Linn.

(Family: Clusiaceae)

Hindi *Mangustan*; Burmese *Mangut.*

Field Identification: Leaves thick, leathery, elliptic-oblong, 15-25 cm, with numerous parallel lateral nerves, intermediate nerves joined by intra-marginal veins. Flowers usually bisexual, 3.75-5 cm across, purple or yellow-red, sepals circular, petals broad-ovate and fleshy. Stamens many, ovary 5-8 celled, stigma thick, 5-8 lobed. Fruit globose, 6.25 cm across, dark purple with white delicious pulp inside. **Description:** An evergreen medium sized tree up to 18 m and more under favourable conditions. **Leaves** thickly leathery, dark green, 15-25 cm, elliptic oblong, with numerous parallel lateral nerves, alternating with shorter intermediate nerves joined by distinct intra-marginal veins. **Flowers** mostly bisexual, 3.75-5 cm across, purple or yellow red in few flowered terminal bunches; sepals circular and petals broad-ovate and fleshy. Stamens many, filaments slender, flat at base and sometimes fused. Ovary 5-8 celled, stigma sessile, thick, 5-8 lobed. **Fruit** globose, 6.25 cm across, dark purple, rind thick, full of yellow resinous juice. **Seeds** 5-8, flat, large embedded in snow white delicious pulp. **Distribution:** Exact origin unknown, but is believed to be the Malay Peninsula. Cultivated in Sri Lanka, Myanmar and in the Nilgiris. **Phenology: Flowers** November-February. **Fruits** May-June. **Miscellaneous:** It is one of the most highly prized fruits of the tropics. The pulp melts in the mouth like ice cream, in flavour something between grape and peach. **Gardening:** Propagation by seeds from ripe fruits sown within 5 days of collection.

11. IRONWOOD TREE
Mesua ferrea Linn.

(Family: Clusiaceae)

Bengali, Hindi *Nagkesar*; Assamese *Nahor*; Gujarati, Marathi *Nagchampa*; Tamil *Nangal*; Telugu *Nagkesara*; Burmese *Gangaw*.

Field Identification: Generally medium-sized, white rose-like flowers 2.5-10 cm across with a central bunch of yellow stamens. Petals 4, larger than sepals, young leaves crimson above, silvery beneath. **Description:** A medium-sized to large, handsome, evergreen tree with a rounded crown, often buttressed. Up to 45 m or more in the Western Ghats, where it sometimes forms part of the emergent layer in tropical rain forests. Elsewhere of moderate size up to 24 m where it occurs in the second storey as in Assam and Arunachal Pradesh. **Bark** grey, afterwards dark brown or cinnamon, peeling off in white flakes exposing a reddish brown surface, fibrous and red inside and exuding an aromatic resin. **Leaves** opposite, decussate, variable, 6.4-17.8 x 1.3-5 cm, lance-shaped, pointed shining, with wax-like bloom beneath. Young leaves deep crimson above and silvery beneath; petiole 0.5-1.3 cm. **Flowers** solitary, generally terminal, 2.5-10.2 cm across, white with a central bunch of yellow stamens, fragrant and resembling a rose. Sepals 4, in 2 pairs, 1.3-1.5 cm, concave, persistent. **Petals** 4, larger than sepals, wedge-shaped, obovate or obcordate, falling off. **Fruit** 2.5-6.4 cm across, ovoid, pointed, 1-celled, 1-4 seeded, supported by persistent sepals. **Seeds** up to 2.5 cm across, dark brown and shining. The magnificent white flowers with a central mass of golden stamens, visited by bees, make the tree what Sir W. Jones called "One of the most beautiful on earth, and the delicious odour of the blossoms justly gave them a place in the quiver of Kamadeva." An outstanding foliage tree; the young leaves are blood red, then pass through shades of pink to dark green. The earliest leaves, emerging from the germinating seeds are dark blue in Tirunelveli and pink in the Silent Valley. There is a belief that blue leaves indicate trees whose heartwood is black, while the pink and pale green ones indicate dark red to reddish-brown heartwood. **Distribution:** Evergreen and semi-evergreen forests of NE. India, Western Ghats, Andamans, Sri Lanka, Bangladesh and Myanmar at 60-1,200 m, sometimes to 1,500 m. **Phenology:** New **leaves** in February in NE. India, from December-February in Western Ghats, Andamans, Sri Lanka, Bangladesh and Myanmar at 60-1,200 m, sometimes 1,500 m. **Flowers** February-April in Western Ghats, February to June Eastern India. **Fruits** July-September in NE. India, May-January Western Ghats. **Miscellaneous:** The fragrant blossoms are used for stuffing pillows in bridal beds. One of the five arrows of Kamadeva, the Indian Cupid, is tipped with mesua wood. A sacred tree, grown around temples. The oily seeds are strung like beads on a piece of split bamboo and when lighted burn like a candle. The Nagas make bows from the wood. Also used for golf club heads, country pipes, as props in Kolar gold mines. The leaves and flowers are used against snake bite. **Gardening:** Grown from seed. They lose viability quickly and therefore are sown immediately. Germination in about 10-14 days. Two-three year old saplings are best for transplanting. It deserves to be widely grown in gardens in areas of good rainfall.

12. POONSPAR TREE, SPAR TREE, POON
Calophyllum polyanthum Wall. ex Choisy

(Family: Clusiaceae)

Syn. *C. elatum;* Bedd., *C. tomentosum* auct. non Wight.

Tamil *Kattupinnai*; Marathi *Nagani*; Malayalam *Kattupunna*.

Field Identification: Poon is easy to recognise provided one can obtain the leaves. This is difficult because of its enormous height. Leaves are therefore picked up from the forest floor. The leathery texture with close parallel nerves at once separates it from other genera. A tall evergreen, with a columnar trunk up to 5 m in girth. The wavy vertically fissured yellow bark is distinctive. Leaves leathery, shining 7.5-12.6 x 3-5 cm with numerous parallel nerves. Flowers white, scented, 1.3-2 cm across. Fruit ovoid, 2 cm long, pointed. **Description:** An evergreen tree of enormous height, 46 m or more and with a columnar trunk 5 m in girth. It attains this giant size in the Western Ghats, being one of the tallest trees of the rain forests. It rivals *Dipterocarpus indicus* where the two are associated. Young shoots, buds and panicles rusty hairy; young branches quadrangular. **Bark** yellowish, with long vertical fissures, a character which helps field identification. Wood red. **Leaves** handsome, leathery, shining, 7.5-12.6 x 3-5 cm, oblong lance-shaped, pointed at the tip and base, lateral nerves fine, parallel, prominent, petioles 1.3-2 cm. **Flowers** white, scented, 1.3-2 cm across in terminal panicles. Sepals 4, the outer circular, shorter than the inner, fringed with fine hairs. **Petals** 4, oblong, obtuse, larger than sepals. **Fruits** ovoid, 2 cm long, pointed. **Distribution:** Evergreen and semi-evergreen forests of the Western Ghats from North Kanara to Kanyakumari and Sri Lanka up to 1,500 m. **Phenology:** February-March. **Fruit** ripens in May-June. In North Kanara the flowers appear in January-February and the fruit ripens in June-July. **Miscellaneous:** It is called the Spar tree because the Arabs used to come to India in search of logs for the masts of their 'dhows'. The price paid was the number of rupees laid edge to edge along the length of the spar. The wood is strong and elastic, useful for bridges, railway sleepers, masts and spars of ships. Only seeds are used for illumination.

13. BALLAGI
Poeciloneuron indicum Bedd.
(Family: Clusiaceae)

Kannada *Ballagi*; Malayalam *Vayal*; Tamil *Pathang*; Trade *Ballagi*.

Fig. 3

Field Identification: A tall tree standing on curved stilt roots. Bark with thick yellow juice. Leaves elliptic 9-25 x 5.8-6.5 cm, leathery, pointed at the tip with closely parallel nerves. Flowers yellowish white, fragrant, 1.9 cm across. **Fruit** egg-shaped, 1-seeded, dehiscent capsule 2.5-3.8 cm in diameter. **Description:** A large evergreen tree up to 36 m and more, crown at first conical, umbrella shaped at maturity, at once recognized in the forest by its massive stilt roots which in old trees are flattened into plank buttresses. **Bark** dark brown to dark grey, with thick yellow juice. **Leaves** opposite, elliptic, 9-25 x 3.8-6.5 cm with a pointed tip, called 'drip tip', an adaptation to drain off rain water, leathery and smooth, with close-set, parallel nerves which are not prominent. **Flowers** bisexual, yellowish-white, about 2 cm across, fragrant, in pyramidally spreading terminal panicles, which are 10-15 cm long. Sepals 4-5. Petals 5-6. **Fruit** egg-shaped, 1-seeded dehiscent capsule 2.5-3.8 cm across. **Seeds** erect. Two varieties of Ballagi are found in Karnataka: one with brown or red heart-wood, called the 'white variety', the other with black heart-wood called the 'black variety'. Apparently, there is no taxonomic difference between the two, though there are differences in habit, habitat etc. The bark of the 'black variety' is darker and the panicles are borne on shorter stalks. **Distribution:** Endemic in the evergreen forests of the Western Ghats from Southern Kanara southwards with breaks in distribution in the 'ghats' of Hassan and Coorg and in parts of Wynaåd. Common at 300-900 m on wind-swept ridges. Typically a species of the west coast tropical evergreen forest. **Phenology:** Leaves shed in the dry season in February-March. Flowers December-March. At the peak of flowering season from January to February, the flowering is so profuse that the crowns of trees are covered with cream-coloured flowers and the air is surcharged with their fragrance. **Fruits** June-August. **Miscellaneous:** It is one of the most important woods of the Western Ghats, long used for power line poles. Also used as railway sleepers.

FIG. 3

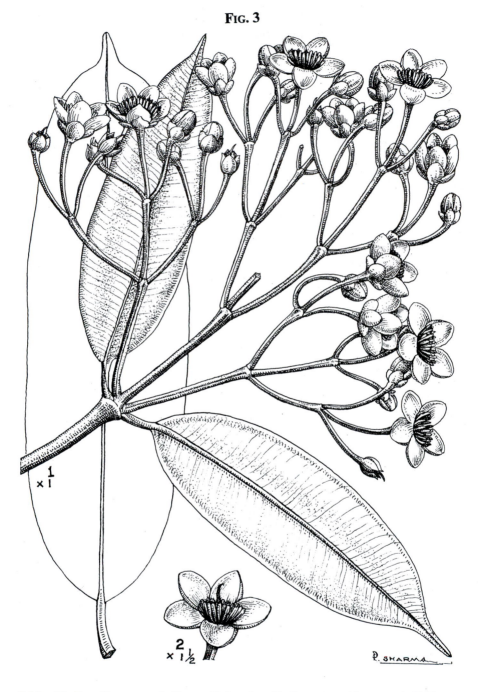

S.No. 13. ***Poeciloneuron indicum***. 1) A twig with flowers and leaves x 1. 2) Open flower x 1½.

14. NEEDLE WOOD
Schima wallichii (DC.) Choisy

(Family: Theaceae)

Hindi *Kanak*; *Makrisal*; Bengali, Nepali *Chilauni*.

Field Identification: A large tree, bark dark grey with vertical clefts. The cut surface when seen under a hand lens shows numerous needle-shaped crystals which cause itching on contact. Leaves oblong or elliptic lance-shaped, 9-24 x 3.5-8 cm with reddish lateral nerves. Flowers white, fragrant, solitary, axillary, 3-5 cm across. Fruit globose, depressed, 1.3-1.8 cm across, covered with warts. Seeds 2-6 in each cell, surrounded by a paper wing. **Description:** A large, handsome, evergreen tree 20-30 m tall. **Bark** thick, soft dark grey, with deep vertical clefts. Blaze red, juicy; the cut surface when examined under a hand-lens reveals numerous needle-shaped crystals of oxalate, which cause intense itching when touched. The Nepali name of this tree is *Chilauni*, which means itching. **Leaves** 9-24 x 3.5-8 cm, oblong or elliptic, lance-shaped, pointed, usually entire, thinly leathery, shining above, hairy beneath, especially along the midrib; lateral nerves reddish, petiole 1.5-2 cm long. **Flowers** white, fragrant, solitary axillary, 3-5 cm across. **Sepals** 5, 0.4-0.5 cm, rounded, densely silky inside. Petals 6, silky hairy towards the base, 1.5-2 cm x 1.3-1.5 cm. **Fruit** a 5-celled capsule, 1.3-1.8 cm in diameter, globose-depressed, covered with warts, supported by a persistent calyx. **Seeds** 2-6 in each cell, flat, greyish brown, surrounded by a papery wing. **Distribution:** From Nepal to Arunachal Pradesh, Assam, Khasi Hills, Manipur, Chittagong hills tracts, Bangladesh, and Upper Myanmar. It is usually found in the foothills and ascends to 1,500 m or more. It attains its best dimension in the 'duars' of W. Bengal in fine quality sal forests. **Phenology:** Evergreen, but sometimes sheds its leaves and becomes almost leafless by the end of March. New **leaves** appear in March-April and are of a delicate pink colour. **Flowers** April-May. **Fruits** November-December. The seeds are light and dispersed by wind. **Miscellaneous:** Parakeets break open the capsules and eat the seeds. Used for railway sleepers, dugouts, constructional purposes and tea chests.

15. GURJAN
Dipterocarpus turbinatus Gaertn.f.

(Family: Dipterocarpaceae)

Hindi, Bengali *Gurjan*; Assam *Tilia gurjan*; Burmese *Kanyin*.
Trade: *Gurjan*.

Fig. 4

Field Identification: Lofty tree with light grey bark. Leaves ovate, drying to coppery brown. Stipules sheathing the apical bud, 5 cm long, when pressed the unopened stipules explode with a 'pop'. The fallen stipules are found on the forest floor in great numbers. Fruits spindle-shaped. **Description:** A lofty evergreen tree, up to 37-46 m, and a girth of 4.6 m with a long cylindrical bole and an elevated crown. **Bark** light grey, longitudinally fissured, cracking in strips. **Leaves** 12.5-30 x 6.3-12.3 cm, leathery, ovate or lance-shaped, margin entire or distantly toothed, with 10-20 pairs of parallel nerves; petiole 2.5-4.3 cm; stipules large up to 5 cm, buff hairy, sheathing the apical bud. The fallen stipules leave behind a prominent scar. **Flowers** 3-3.5 cm, white or pinkish; petals softly hairy, 2.5-3 cm. **Fruit** body without ribs, spindle shaped, with linear marks on surface on maturing; fruit-belly 2-3 x 3-3.8 cm, smooth, ovoid or ellipsoid, very shortly stalked, wings two, terminal, 11.4-17.8 x 2.5-3.8 cm, with three nerves in the lower half, middle prominent, the two veins disappearing about half way up the wings. **Distribution:** Cachar and Silchar in Assam, mostly found in the Surma Valley, and in Manipur, Tripura, Meghalaya (Garo Hills), Mizoram, extending into Myanmar, Chittagong Hills (Bangladesh). Mostly on low undulating hills below 300 m. Its occurrence in the Andamans is doubtful. **Phenology:** Old leaves are generally shed early in the hot season. **Flowers** January to March, occasionally up to April. **Fruit** ripen and fall from May to June. When the fruits drop they rotate in the air like the propeller of an aircraft, so that they are blown far and wide. The seeds have a poor viability and are therefore sown within 7-10 days of collection. **Miscellaneous:** A valuable commercial timber commonly used for boat building, railway sleepers, planking, flooring and dugout canoes. The timber is slightly stronger and harder than teak. **Associates:** *Aquilaria agallocha, Artocarpus chaplasha, Mesua ferrea, Shorea robusta, Tetrameles nudiflora,* etc. which are described in this book.

FIG. 4

S.No. 15. **Dipterocarpus turbinatus**. 1) A branch with leathery glossy leaves and pinkish flowers. 2) Spindle-shaped fruits with papery wings. Both x 1.

16. SAL
Shorea robusta Gaertn. f.

(Family: Dipterocarpaceae)

Hindi Trade *Sal*

Field Identification: Gregarious, with shining leathery leaves 10-20 x 6-12 cm, ovate oblong with 12-15 lateral nerves. Flowers in panicles. Pale yellow, fragrant, covering the entire tree when leafless. Petals 1.3 cm long. Fruit 1.25 cm, leathery with five terminal spoon-shaped wings 6.3 cm long, brown when dry. **Description:** A large, semi-deciduous, gregarious tree of monsoon forests, with shining foliage, covering large areas. Generally 13-32 m tall with a girth of 1.5-2 m. Trees of exceptional height and girth are known from Haldwani, Uttar Pradesh, a protected tree measured 51 m in height and g.b.h. of 2.64 m. A girth of 6.40 m was recorded from Tripura. **Bark** dark brown, with longitudinal fissures. **Leaves** leathery, ovate-oblong, 10-20 x 6-12 cm, with 12-15 lateral nerves. Young leaves are red or coppery coloured, an adaptation against intense light; stipules 0.8 cm, sickle-shaped, hairy. Panicles 12.5-22.5 cm long, clothed with pale velvety hair. **Flowers** fragrant, short stalked, petals *c*. 1.3 cm long, pale yellow, softly hairy, tapering upwards; stamens up to 50, much shorter than petals, connective awl-shaped, bearded, minutely trifid at apex. Ovary 3-celled. **Fruit** 1.25 cm long, leathery, indehiscent, 1-seeded, with 5 terminal wings 6.3 cm long, brown when dry, spoon-shaped with 10-12 parallel nerves. **Distribution:** In the north in the sub-Himalayan tract from the Kangra Valley to Arunachal Pradesh through Garhwal, Kumaon, Nepal, Sikkim and Bhutan. South of the sub-Himalaya in Haryana, Uttar Pradesh, Bihar, Orissa, Bengal, Assam, Meghalaya, Tripura, Madhya Pradesh and Andhra Pradesh. The altitudinal range is about 10 m in Tripura to more than 1,500 m in the valleys of some rivers of Kumaon and Garhwal. The highest recorded is 1,680 m from Pithoragarh. **Phenology:** The leaves turn yellow and commence falling from January to March. In dry areas, the trees are leafless for a short time. The young leaves are reddish and shining. New leaves appear from February-May. **Flowers** in March, when leafless. The trees are covered with masses of blooms and are a striking sight. **Fruit** ripens in June and the seeds germinate immediately. **Miscellaneous:** Seedlings are browsed by deer, gaur and nilgai. Elephants damage young plantations. Sal ranks second to teak in importance. It is resistant to termites and is in very great demand as railway sleepers and telephone poles. It is an important furniture wood. The stem exudes a gum called 'Sal dammar' used in marine yards for caulking boats, also in shoe polish. Sal butter extracted from seeds is used in chocolates and also as a cooking agent. The leaves are used for rolling home-made cigars by the adivasis (tribals) of Central India.

17. INDIAN COPAL TREE
Vateria indica L.

(Family: Dipterocarpaceae)

Trade name *White Dhup*; Kannada *Guli;* Coorg *Bilidupa;* Tamil *Vellei*; Malayalam *Payia*.

Field Identification: Trunk cylindrical with smooth, whitish grey bark. Young leaves red, mature leaves 10-20.3 x 5-10 cm, ovate, shortly pointed, with 14 pairs of nerves. Flowers 2 cm, white fragrant and drooping. Stamens 40-50. Fruit ovoid, 3 valved, leathery, fleshy, 5-6.2 cm with remains of the calyx. Seed large. **Description:** A large handsome evergreen tree with smooth whitish grey bark blotched with white and green, bitter and acrid in taste. **Inflorescence** clothed with mealy-grey star-shaped hairs. The young leaves are bright red to copper coloured, on maturity they are leathery, 10-20.3 x 5-10 cm, ovate shortly pointed with 14 pairs of nerves, prominent beneath. **Flowers** 2 cm across, white, fragrant, dropping in large panicles. Petals spreading; stamens 40-50, ovary 3-celled. **Fruit** a large 5-valved capsule, supported by the remains of the calyx, fleshy, filled with fat (Piney tallow). **Distribution:** South-west India from north Kanara (Karnataka) to Tirunelveli and Kanyakumari (Tamil Nadu) to Sri Lanka. In the ghats in Coorg, trees over 5 m girth are recorded. From 60-760 m along stream banks, valleys and coastal areas. **Phenology:** An evergreen without a pronounced leaf shedding season, maximum leaf-fall during the hot season in March. The new copper coloured leaves appear after March. **Flowers** January to March. Fruit ripens from May-July. Viviparous (germinating on the tree). **Miscellaneous:** Tea chests, coffins, flooring, ammunition boxes, as railway sleepers after treatment and oars for sea going vessels. It yields a resin, called white dammar, or Indian copal. The resin, when mixed with coconut oil, makes an excellent varnish. Kernels yield a fat called piney tallow which is used to adulterate ghee. **Gardening and propagation:** A handsome tree for parks and avenues in South India. The bright red young leaves appear during the cold season, equally spectacular when the tree is full of fragrant white flowers from January to March. The petals cover the ground during March and April with masses of yellow anthers. Direct sowing of fresh ripe seeds gives good results. The tree loves moisture.

18. THINGAM
Hopea ponga (Dennst.) Mabberley

(Family: Dipterocarpaceae)

Syn. *H. wightiana* Wall. ex Wight & Arn.

Kannada *Hiribogi*; Malayalam *Pongu*; Marathi *Kavsi*; Tamil *Ilapongu*; Trade name *Thingam*.

Field Identification: A tree with a fluted stem. Fruit crimson, 2-winged. Nut 1.3 cm, wing 5-7 x 1.3-1.5 cm. Bark smooth, brown, peeling in large rectangular plates, inner bark white or yellowish. **Description:** Evergreen tree 35-40 m high with a fluted and tapering stem. **Bark** brown, smooth, peeling off in rectangular pieces. Young **branches** and petioles rusty hairy. **Leaves** oblong, lance-shaped, leathery, 12.5-20 x 5-7.5 cm, rounded at the base, nerves 9-12 pairs, prominent beneath, midrib strong. The inflorescence is very distinctive in that the flowers are secured on the branches of the panicle. **Flowers** pinkish-white, 0.85 cm across, 3 together on 15 cm long panicles, petals 0.5 cm long, ciliate on the margins, filaments 10, dilated at the base. **Fruit** an ovoid nut, 1.3 cm long, surrounded by the remains of the calyx, 2 of the lobes of which enlarge into two crimson wings, 5-7 x 1.3-1.5 cm. The flower buds are often attacked by an insect and transformed into round, solid spiny galls which is a distinguishing feature of Thingam. **Distribution:** SW. India in semi-evergreen forests in 'ghats', often gregarious in forests fringing river banks. Goa, Konkan, Karnataka, Tamil Nadu and Kerala mostly at 320-800 m. **Phenology:** **Flowers** March to April. **Fruits** May, June or July. The trees are very handsome when covered with crimson fruit in the hot season. **Miscellaneous:** Beams, rafters, cart wheels, shuttles and textile machine accessories.

19. INDIAN TULIP TREE, UMBRELLA TREE, PORTIA TREE
Thespesia populnea (L.) Soland. ex Correa

(Family: Malvaceae)

Marathi *Bhendi;* Hindi, Gujarati, Bengali *Paras Pipal*; Tamil *Poovarasam Kallai*; Malayalam *Poovarasu*.

Field Identification: Small tree with an umbrella-shaped crown. At once recognized by its tulip-like yellow flowers with a purple centre, its heart-shaped poplar-like leaves and the black turban-shaped capsules. **Description:** A small evergreen tree with an umbrella-shaped crown, up to 12 m high. **Bark** grey, inner bark fibrous. **Leaves** poplar-like, 8-15 x 6-10 cm, heart shaped, entire, pointed at the tip; petiole 2.5-10 cm; stipules awl-shaped, deciduous. **Flowers** axillary, solitary or two together, bisexual, bell-shaped, 5-7.5 cm across, pedicels 5-7.5 cm long. Calyx cup-shaped, covered with minute, stalked scales. **Petals** 5, yellow with a crimson or deep chocolate centre, resembling yellow crepe paper. **Fruit** a globose capsule, 2.5-4 cm across, covered with minute stalked scales, calyx persistent, black when ripe. Seeds *c.* 1 cm long, ovoid. **Distribution:** Common on the coast of South India, Andamans, Bangladesh, Myanmar and Sri Lanka. Cultivated as an avenue tree in Chennai, Mumbai, etc., as far inland as Chandigarh. **Phenology:** Although evergreen in moist areas, it sheds many leaves in February in less moist areas. The yellow withering leaves resemble flowers, giving the impression from a distance that the tree is in full bloom. **Flowers** in the cold season, also throughout the year. **Fruits** April-May. **Miscellaneous:** Propagated from seeds and cuttings. It is mostly planted in gardens in Chennai and Mumbai but is capable of growing everywhere except in the hills. It is a shady roadside tree and is also suitable for small gardens. It is used for boat building. The fibres of the bark are used in cordage. It is suitable for avenues as its umbrella shaped crown provides shade.

20. VEDI PILA
Cullenia exarillata Robyns

(Family: Bombacaceae)

Tamil *Vedipila*; Malayalam *Karayani*.

Fig. 5

Field Identification: A very tall tree of the Western Ghats, forming the top storey. Leaves oblong or oblong lance-shaped with very close parallel nerves and undersurface with silvery or orange coloured peltate scales. Flowers rusty red, borne on the old wood. Fruit capsular, 10-12.5 cm long, covered with long spines. **Description:** A very tall tree, wood moderately soft. Young shoots covered with shield-shaped stalked scales. **Leaves** oblong, lance-shaped with close parallel nerves, entire, with a tail-like pointed tip, smooth above, surfaced beneath with silvery or orange-coloured, stalked scales. **Flowers** rusty red, in large clusters borne on the old wood. Secondary bracts (bracteoles) fused into a tube splitting irregularly into 3-5 lobes. Calyx tubular, 5-lobed; bracteoles and the calyx completely covered with large shield-like, stalked scales. **Petals** 0. Staminal tube sticking out, divided into 5 finger-like segments, each segment bearing along its length numerous minute anthers in globose clusters. Ovary 5-celled, style longer than stamens, stigma globose. **Fruit** 10-12.5 cm long, covered with long prickles, valves 3-5. **Seeds** few, 3.3 cm long, hard, shining, nearly covered by a large fleshy

FIG. 5

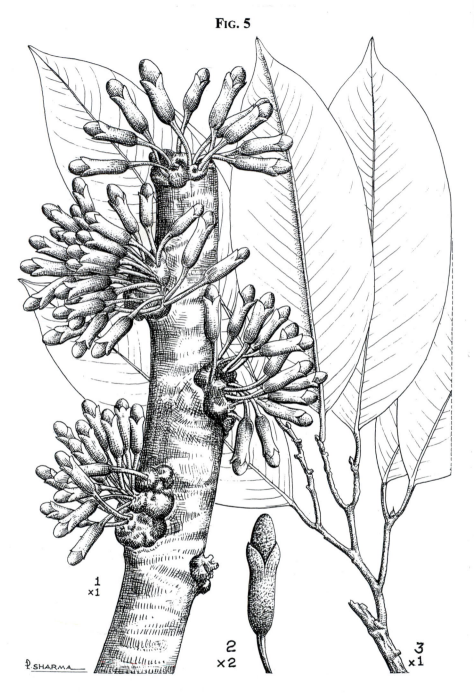

S.No. 20. *Cullenia exarillata*. 1) Flowering branch x 1. 2) Flowers x 2. 3) Leaves x 1. (From herbarium sheet 105238 in DD).

white pulpy covering. Cotyledons fleshy. **Distribution:** Western Ghats up to 1,200 m, from Coorg southward through Kerala to Kanyakumari Dist., Tamil Nadu, forming the top storey in moist regions from 600-1,300 m. **Phenology: Flowers** in the hot season. **Fruits** about June. **Miscellaneous:** A major food source of the rare and endangered lion-tailed macaque (*Macaca silenus*).

21. RED SILK COTTON
Bombax ceiba L.

(Family: Bombacaceae)

Plate 2

Hindi, Bengali *Semul*; Gujarati *Sanar*; Tamil, Malayalam *Mulilavu*; Marathi *Shevri*; Burmese *Leptan*.

Field Identification: Lofty buttressed tree with tiers of whorled branches covered with conical prickles when young. Leaf with 5-7 lance-shaped leaflets arranged like the fingers of a hand. Flowers large, brilliant crimson, petals 7.5 x 2.5 cm, stamens 75 in bundles of six, one bundle around the style. Fruit a 15 cm long pod with many small seeds embedded in silky floss. **Description:** A lofty deciduous tree up to 40 m with a buttressed base. **Bark** covered with conical prickles when the tree is young. Characterized by tiers of whorled branches. The **leaf** is composed of lance-shaped leaflets arranged like the fingers of a hand, common petiole 15-30 cm long; leaflets 5-7, shortly stalked, ovate, pointed at the tip, entire, 10-20 x 3.2-5.5 cm. **Flowers** large, brilliant crimson, rarely yellow or orange in Assam or rarely white in the Western Ghats, fascicled at or near the ends of the branches. Calyx cup-shaped, usually 3-lobed, smooth outside, white-silky within. **Petals** oblong, scarlet, white-hairy outside, 7.5 x 2.5 cm, obtuse, with close parallel veins. Stamens *c*. 75 in 6 bundles, one of which is around the style; flattened, the innermost forked, half as long as the petals. Ovary 5-celled, style 5-lobed at the apex. **Fruit** ovoid, 5-angled, short stalked, downy 10-15 cm long, black when ripe. **Seeds** small, innumerable, embedded in silky cotton dispersed by wind. **Distribution:** Widespread in the Subcontinent, in the Himalaya up to 1300 m, also in Myanmar and Sri Lanka and as far as northern Australia. **Phenology:** Leafless from December until March. **Flowers** brilliant crimson, close-set, burst open from the dark buds during January to March. **Fruits** split open on the tree in April-May and disgorge quantities of silky cotton in which the small seeds are embedded. **Miscellaneous:** The floss is used for stuffing pillows and is excellent for making surgical dressings. The wood is used for making match boxes and packing cases. A fast growing species. *B. insigne* Wall. of South India, Andamans and Myanmar is similar to the above and differs in having 400 stamens and larger petals up to 12.5 cm long.

22. WHITE SILK-COTTON, TRUE KAPOK
Ceiba pentandra (L.) Gaertn.

(Family: Bombacaceae)

Tamil *Ilavam*

Field Identification: A small buttressed tree with whorled branches. Young stems bear conical prickles. Leaf with 5-8 lance-shaped leaflets arising from a long stalk. Flowers dirty white with a milky smell, much smaller than 'semul'. Stamens 5. Fruits 7.5-12.5 x 5 cm, 5-celled, seeds in tufts of silky hair. **Description:** A small to medium-sized tree, large and lofty in the Amazon from where it was introduced long ago. It is now common in South India, Sri Lanka and Myanmar. Young stems bear conical prickles, the branches arise in whorls and the adult trees are buttressed. The **leaf** has 5 to 8 lance-shaped pointed leaflets 7.5-10 x 2.5 cm arising from a long stalk. **Flowers** dirty white with a milky smell and much smaller than those of the red silk cotton. Calyx bell-shaped with 5 obtuse teeth, persistent. **Petals** twice to three times the length of calyx. The specific name which means 'five stamens' helps to distinguish it from the red silk cotton which has 75 stamens. In kapok the stamens are fused at the base into a fleshy tube and are shorter than the petals. Ovary conical, style stout, exceeding the stamens; stigma 5-lobed. **Capsule** ovoid, oblong, 7.5-12.5 x 5 cm, 5-valved, many seeded, **seeds** enclosed in separate woolly balls. **Distribution:** Introduced from the Amazon, common in South India, Andamans, Sri Lanka and Myanmar. **Phenology:** Deciduous in the cold season, new leaves appear in December **Flowers** December-January. **Fruits** March-April. **Miscellaneous:** The pale yellow floss is world-famous for use in life-buoys and life belts. The red silk cotton is also loosely called kapok, and is passed off as true kapok by unscrupulous traders. The ease with which cuttings of kapok take root has led to its extensive use as a hedge tree by farmers in South India.

23. CIVET FRUIT, DURIAN
Durio zibethinus Murr.

(Family: Bombacaceae)

Fig. 6

Field Identification: Heavily buttressed tree with drooping leaves, golden hairy beneath. Fruits are like a football 15-25 cm covered with thick spikes. **Description:** A lofty evergreen tree *c.* 23 m tall with a massive trunk and heavily buttressed base. Young trees conical, on maturity with a spreading crown and gnarled and crooked branches. **Bark** warm brown, rough and flaky. **Leaves** drooping, dull brownish green, golden hairy beneath, leathery, oblong lance-shaped 11.5 x 3.5

FIG**. 6**

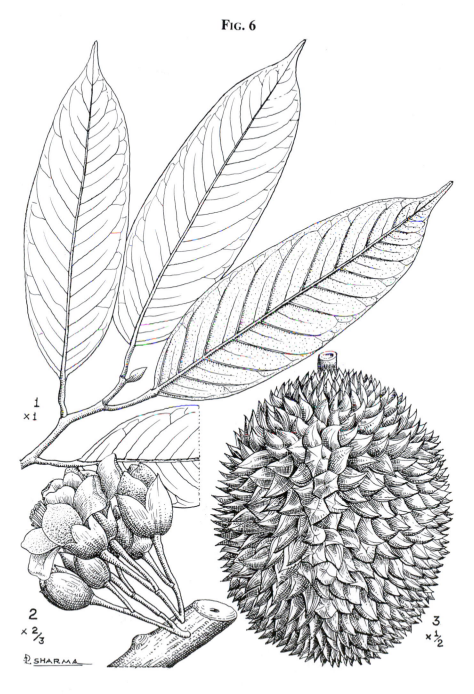

1
× 1

2
× 2/3

𝒫 SHARMA

3
× 1/2

S.No. 23. **Durio zibethinus**. 1) Leaves x 1. 2) Flowers x $\frac{2}{3}$. 3) Fruit x $\frac{1}{2}$.

cm. **Flowers** cream-yellow c. 4 cm across arising in tufts, calyx bell-shaped, 5-lobed, petals 3; staminal column divided into many filaments in 4-6 groups, ovary 5-celled, style long with a head-shaped stigma. **Fruit** sub-globose, 15-25 cm long with a spiny wood cover as in jack fruit. **Seeds** are embedded in a cream-coloured edible pulp. The fruit is not fully ripe until it has dropped from the tree: it then begins to split into 5 pieces. **Distribution:** A Malayan tree long under cultivation in the Nilgiris, on the west coast, Sri Lanka and Myanmar at low elevations. **Phenology: Flowers** March-April. The flowers open between 2 and 3 pm and fall off at 2 am apparently self pollinated or pollinated by bats. Bees are seen on the flowers in the afternoon. **Fruit** ripens July-September. **Miscellaneous:** Durian is regarded as the most famous tree of the East. The seeds are embedded in custard-like pulp which, mixed with cream and sugar, has a delicious taste but a putrid smell. A taste for durian is quickly acquired. Seeds are eaten roasted like chestnuts. A red-flowered variety *roseiflorus* has the best taste. It is one of the most beautiful flowering trees. **Propagation:** By seeds sown soon after collection from ripe fruits. Seedlings are transplanted at a spacing of 8-11 m. Propagation by inarching is possible and selected types are multiplied by budding. **Etymology:** From the Italian *Zibetto* or strong smelling.

24. BAOBAB, MONKEY BREAD TREE
Adansonia digitata L.

(Family: Bombacaceae)

Hindi *Gorakh imli*; Gujarati *Gorakh amli*; Marathi *Gorakh chinch*

Fig. 7

Field Identification: A large tree with gigantic girth of 12 m or more, out of proportion to its height. Trunk bottle-shaped. Flowers white, 5-8 cm across, long pendulous. Fruit 15-21 cm x 4-5.8 cm, yellowish felted outside. **Description:** A large tree with a bulbous bottle-shaped trunk which startles observers by its great girth of 12 m or more. It is 12-18 m high, with thick branches. **Bark** whitish, shining sometimes purplish, inner bark fibrous. The leaf is composed of 5 lance-shaped leaflets arranged like the fingers of a hand and borne on a long petiole, leaflets obovate (i.e. terminal half broader than lower half), pointed, very short stalked, up to 12 x 5 cm broad, entire or toothed, with star-shaped hair beneath or smooth. **Flowers:** large 12-15 cm across, white, hanging on long stalks, calyx 5-lobed, hairy on both sides, petals covered with bristles on the outside. Stamens numerous. **Fruit** ellipsoid to globose 15-21 x 4-5.8 cm, hanging on long stalks, yellowish felted outside. **Seeds** 10 x 8 mm, kidney-shaped or circular, embedded

42

Fig. 7

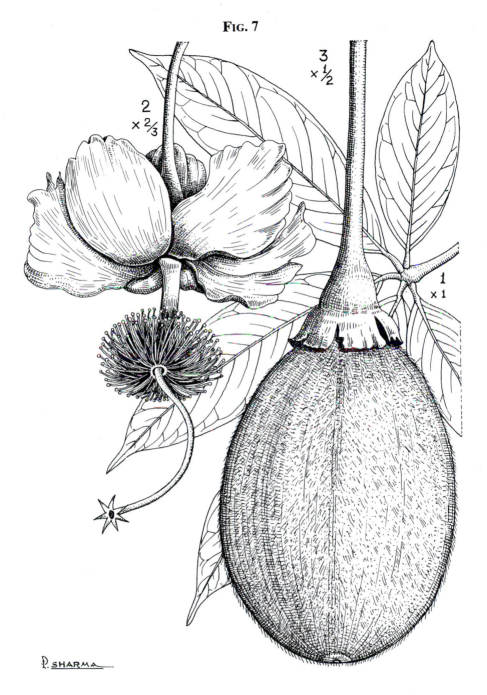

2
× 2/3

3
× 1/2

1
× 1

P. SHARMA

S.No. 24. *Adansonia digitata*. 1) Leaves x 1. 2) Flowers x $\frac{2}{3}$. 3) Fruit x $\frac{1}{2}$.

in dry acid pulp. **Distribution:** Common along the west coast and also in Sri Lanka. Supposed to have been originally introduced by Arabs centuries ago, but probably was also brought by the Portugese from Tropical E. Africa where it is endemic and is one of the longest lived of African trees. **Phenology:** Deciduous, dropping its leaves early in the dry season. **Flowers** May-June. **Fruit** ripens in January and February. **Miscellaneous:** Baobabs live to an enormous age; over 1000 years has been measured by radiocarbon dating and much greater age by less precise techniques. They do not continue to grow in size year by year, and big trees may even shrink. This is attributed to drought. Monkeys relish the fruit and hence the popular name, monkey bread tree. In Africa it is reported to have the property of preserving human corpses. All parts of the tree are useful, providing timber, fibres, leaves as a vegetable and acidic seed as an important source of vitamin C. **Gardening:** This fast growing tree is worthy of introduction in parks and botanical gardens in tropical forest areas of the Subcontinent. Commonly propagated by seed and planted singly in parks because of its enormous size. **Etymology:** *Adansonia* is named after Adanson, a French botanist and *digitata* refers to the finger-like arrangement of the leaflets.

25. KARAYA

Sterculia urens Roxb.

(Family: Sterculiaceae)

Hindi *Kulu*; Gujarati, Marathi *Karai*; Tamil *Kavalam*; Telugu *Kavili*.

Fig. 8 & Plate 3

Field Identification: Moderate sized, with an irregular gnarled appearance when leafless during the winter. Bark smooth, white or greenish-grey, scaling off in thin papery plates like birch (bhojpatra). Leaves at ends of branches 5-7 lobed, hairy beneath, lobes oblong or ovate-oblong, toothed or lobed, blade 17.5-20 cm across. Flowers yellow, 0.6-1.2 cm. Fruit of 2-5 follicles, bright red, 3.7-7.4 cm long, clothed with stiff stinging bristles. Seeds 3-6 in each follicle, dark brown. Flowers and shoots have a bad odour. **Description:** A moderate-sized tree, with an irregular, gnarled short trunk. **Bark** smooth, shining, greenish white, the thin papery bark scaling off like birch bark. Wood soft, reddish brown with an unpleasant smell. **Leaves** at ends of branches, palmately 5-lobed pointed at the tip, heart shaped; blade 17.5-20 cm across; petioles 15-25 cm long. **Flowers** yellow, 0.6-1.2 cm across, glandular-hairy, in bunches at the end of the branches when the tree is leafless, male and female or bisexual flowers mixed; calyx 0.6 cm long, bell shaped. Male flowers filaments 10, alternately long and short. Bisexual flowers carpels

Fig. 8

1 x1

2 x ½

3 x ½

P. SHARMA

S.No. 25 *Sterculia urens*. 1) Flowering shoot x 1. 2) Leaf x ½. 3) Fruit with seeds x ½.

usually 5, on a short stalk, style short, stigmas 5. **Fruit** of 5 egg-shaped, leathery, red carpels, covered with stiff, stinging bristles. **Seeds** black or dark-brown, 3-6 in each carpel. **Distribution:** Tropical Himalaya from Ganga eastwards, E. and W. Peninsulas, South India, Sri Lanka and Myanmar. Also near the coast on rocky soil in the Konkan and North Kanara. **Phenology:** Leafless during the cold weather. **Flowers** January. **Fruits** April-May. **Miscellaneous:** Used for toys, musical instruments and canoes. The bark exudes a white gum called "karaya" gum used in cosmetics and for thickening ice-cream. The seeds are roasted and eaten by the Gonds of Madhya Pradesh. A useful fibre for making ropes and coarse cloth is obtained from the bark. This cloth was made into garments in the past in North India. **Etymology:** *Sterculia* is from Sterculius, a God of Roman mythology, derived from *Stercus*, dung. The Romans deified objects they disliked. The flowers and leaves of some species are foul smelling.

26. BONFIRE TREE, THE COLOURED STERCULIA
Firmiana colorata (Roxb.) R. Brown

(Family: Sterculiaceae)

Plate 4

Hindi *Bodula*; Bengali *Mula*; Marathi *Kaushi*; Tamil *Malam herutti*; Malayalam *Malam*; Burmese *Wetshaw*.

Field Identification: Bark ash-coloured. Leaves at ends of branches, shallowly 3 or 5 lobed, lobes triangular, pointed, blade 15-30 cm. Leafless when in bloom. Flowers coral-red 2 cm long. Fruits of 2-5 leaf-like valves pink turning red. **Description:** A tree with a straight, sometimes fluted trunk with ash coloured bark. **Leaves** at ends of branches on long slender stalks up to 30 cm, 3 or 5 lobed, heart shaped, lobes triangular, pointed, blade 15-30 cm across. Petiole 10-30 cm. Young leaves and shoots downy. **Flowers** in short dense clusters at ends of branches, bright coral or orange-red 2.5 cm long. Floral region covered with soft velvety star-shaped hairs; calyx funnel-shaped, 2 cm long, the column of united stamens protruding and bearing about 30 yellow anthers. Petals 0. Stigma short, recurved. Flowers deep red inside. **Fruit** conspicuous, composed of 2 to 5 leaf-like membranous valves, pink turning red on a common stalk. **Seeds** one or two borne on the edges. **Distribution:** Satpuras up to 1,100 m, Mt. Abu; Western Ghats from South Kanara to Kerala, Konkan and the Deccan, North Circars, Andamans, Bangladesh, Myanmar and Sri Lanka. **Phenology:** Leaf shedding November onwards, leafless January to April. New leaves April-May. **Flowers** March to May. **Fruits** May to June. The bonfire tree is conspicuous and a brilliant sight when in bloom. In the Deccan and the Western Ghats where the tree is common it dots the hill sides and ravines like masses of flaming red coral. **Miscellaneous:** Ideal for parks and gardens. The Singhalese admire it greatly and have composed songs in praise of its beautiful flowers. Grows from seeds. **Etymology**: *Firmiana* commemorates Joseph von Firman (1716-1782) a patron of the Padua botanical garden.

27. *Sloanea sterculiacea* var. *assamica* (Benth.) Coode

(Family: Elaeocarpaceae)

Syn. *Echinocarpus assamicus* Benth.

Assamese *Phul Hingri.*

Fig. 9

Field Identification: Moderately large, buttressed deciduous tree. Fruits jet black on maturity, round, 5 cm across densely studded with 2.5 cm long prickles. **Description:** A medium sized deciduous tree up to 18 m in height, often buttressed at the base. Leaves 22-30 cm long, elliptic, abruptly pointed, with parallel secondary nerves, leaf margin smooth, leaf stalks 2.5-5 cm long. **Flowers** bisexual, symmetrical, sepals 4-5, petals 4-5, variously cut. Stamens numerous, free on a thick disc. Anthers with linear opening by a terminal pore. **Fruit** characteristic, round, 5 cm across, densely studded with 2.5 cm long prickles, jet black on maturity, a 3-valved capsule. It is similar to the fruit of the edible chestnut *Castanea sativa* which can be distinguished by the sharply toothed leaves and by the jet black fruit of the former. **Distribution:** Evergreen climax forests (hills and plains) of Eastern India, Assam, Arunachal Pradesh and Sikkim, occurring more or less gregariously on river banks. **Phenology: Leafless** in September-October. **Flowers** October-November. **Fruit** ripens in March-April. **Miscellaneous:** The wood is a pale cream colour, light and soft and is used for tea boxes and planking.

28. UTRASUM BEAD TREE, RUDRAKSH
Elaeocarpus augustifolius Bl.

(Family: Elaeocarpaceae)

Syn. *E. ganitrus* Roxb. ex G.Don, *E. sphaericus* (Gaertn.) K. Schum

Plate 5

Hindi, Nepali *Rudraksh*

Field Identification: Usually buttressed medium sized tree. Leaves lance-shaped 7.5-15 x 2.5-5 cm, obscurely and minutely saw-toothed. Side nerves 10-15 pairs.

FIG. 9

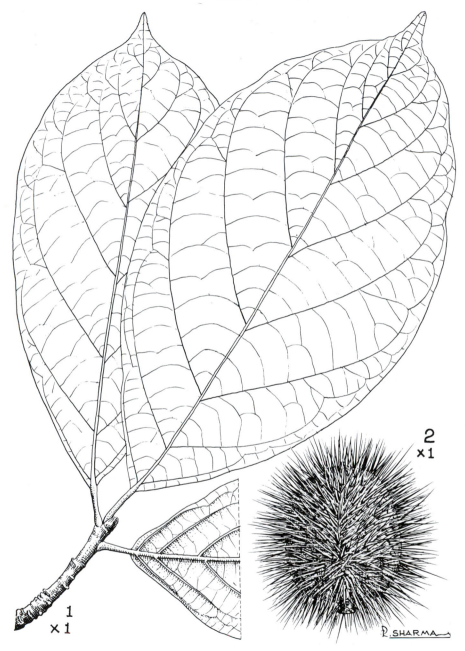

S.No. 27. **Sloanea sterculiacea** var. **assamica**. Leaves and spiny jet black fruit x 1.
(From *K C Sahni* 5073. Arunachal Pradesh).

Some old leaves turn red before falling. Flowers white, 1.25 cm across, petals cut into finger-like lobes. The longer anther valve with 1 or 2 white bristles. Fruit 2-2.5 cm wide, round, bright metallic blue, enclosing an elegantly tubercled stone with 5 grooves. **Description:** A medium sized to large, partly deciduous tree, up to 18 m high, usually buttressed. **Leaves** 7.5-15 x 2.5-5 cm, lance-shaped, slightly toothed, with 10-15 pairs of side nerves, petiole 0.6-1.25 cm long. Old leaves often turn red before falling. **Flowers** 1.25 cm across, white, on sprays 5-7.5 cm long, mostly on twigs behind the leaves. Sepals 5, 0.5-0.8 cm long, petals 5, a little longer than sepals and cut into narrow pointed lobes. Receptacle short, fleshy, wrinkled and hairy. Anthers nearly stalkless, 25-35, the longer valve with 1 or 2 white bristles. **Fruit** 2-2.5 cm wide, round, bright metallic blue, pulp green, enclosing an elegantly tubercled hard brown stone, marked with 5 vertical grooves, 4-5 celled, often one seeded. **Distribution:** Bihar, Bengal, Arunachal Pradesh, Madhya Pradesh, Maharashtra, Nepal and Bangladesh in evergreen forests. **Phenology:** Partly deciduous, some old leaves withering and turning scarlet in autumn. **Flowers** May-June. **Fruits** November-December. **Miscellaneous:** The tree is sacred and popular for ceremonial planting in gardens in India and Nepal. The tubercled 5 grooved stones are much prized; unusual stones with few or more than 5 cells fetch high prices. They are cleaned, polished and used as beads. They look elegant when set in gold in bracelets.

29. BAEL TREE, GOLDEN APPLE
Aegle marmelos Correa

(Family: Rutaceae)

Hindi, Bengali, Marathi *Bael*; Gujarati *Bili*; Telugu *Muredu*; Tamil, Malayalam *Vilvan*; Burmese *Okshit*.

Fig. 10

Field Identification: A spinous tree with gland-dotted trifoliate leaves. Spines 1-3 cm. Leaflets 5-10 x 2.5-6.3 cm, ovate lance-shaped, margins round toothed. Flowers greenish white, honey scented, 2.5 cm across. Fruits 5-15 cm across, round, yellowish woody. Seeds numerous, embedded in orange coloured pulp. **Description:** A small to medium-sized deciduous tree, armed with 1-3 cm long spines. Strikingly handsome on account of its bright green foliage and yellow fruits. **Bark** dark grey, corky, wood yellowish-white, strongly aromatic. **Leaves** alternate, 3-foliate, rarely 5-foliate, leaflets 5-10 x 2.5-6.3 cm, ovate, lance-shaped, gland dotted, margin round toothed. **Flowers** greenish white, honey scented, 2.5 cm across bisexual, petals 4-5 thick, oblong, gland dotted. Stamens numerous, filaments short. **Fruit** 5-15 cm, round, grey or yellowish with a woody shell.

49

Fig. 10

S.No. 29. *Aegle marmelos*. Flower and fruits x 1.

Seeds numerous, flat, clothed with fibrous hair, embedded in a thick sweet, orange-coloured aromatic pulp. **Distribution:** Sub-Himalayan foothills up to 1,200 m from Jhelum (Pakistan) through Jammu and Kashmir to Myanmar, Bangladesh and South India. **Phenology**: Sheds its leaves in April-May for a short period. New leaves April-May. **Flowers** March-July. They cover the entire tree and are a great attraction for honey bees. **Fruit** ripens from January to June, turning yellow and remains on the tree up to July before falling. **Miscellaneous:** The fruit is broken open by deer, pigs, monkeys and possibly bears which eat the pulp. In this way the seeds get dispersed. The sweet nutritious pulp is drunk as a *sherbet*. The leaves are good fodder and the wood is used in agricultural implements. The twigs are used as tooth brushes or chew sticks. The tree is sacred to the Hindus. **Gardening:** Seeds are collected from freshly plucked fruit, not from fallen fruit. They lose viability early. Sown in April-May and are ready for planting in the second or third season.

30. ELEPHANT APPLE
Limonia elephantum (Correa) Panigrahi

(Family: Rutaceae)

Syn. *Schinus limonia* L. *Feronia elephantum* Correa

Fig. 11

Hindi *Bilin, Katbel*; Marathi *Kawoth*; Tamil *Bilva*; Burmese *Thibin*.

Field Identification: A small spinous tree, spines 2.5-5 cm polished at the tips. Leaves 15-25 cm, leaflets 5-7, stalks winged. Fruit round, woody, 7.5 cm across, seeds embedded in brown pulp. **Description**: A small to medium sized deciduous tree up to 9 m tall and a symmetrical crown with thorny branches. **Bark** dark grey, rough and furrowed; spines 2.5-5 cm long, stout and polished at the tips. **Leaves** with 5-7 opposite leaflets; leaflets 2.5-5 x 1.2-2.5 cm, round tipped with a wedge-shaped base. **Flowers** numerous, 0.8 cm across, pale green tinged with red, male and female flowers in the same hairy panicle. Petals 5-6, elliptic-oblong, spreading or bent downwards. **Fruit** round, about the size of a tennis ball or slightly larger, with woody rough covering, containing dark brown subacidic pulp. **Seeds** many, oblong embedded in the edible pulp. **Distribution:** Indigenous in South India and Sri Lanka, cultivated elsewhere and in Myanmar. **Phenology:** Leaves are shed in January for a short period. **Flowers** February-March. **Fruit** ripens October-November. The tree does not fruit till the fifteenth year. The dispersal is aided by animals, apparently elephants seeking the edible fruit. **Miscellaneous:** The leaves

FIG. 11

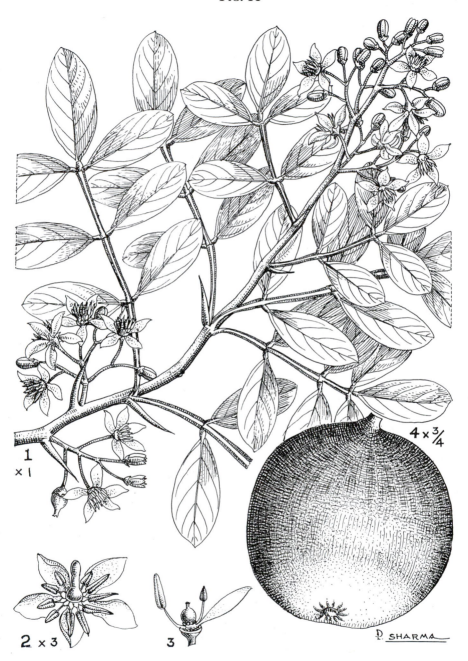

S.No. 30. *Limonia elephantum*: 1) Leaves x 1. 2) Bisexual flower x 3. 3) Male flower (part) x $\frac{3}{4}$. 4) Fruit x $\frac{3}{4}$.

are fodder for sheep and goats. The wood is used in ornamental carvings, and agricultural implements. The transparent gum is used for making water colours. The pulp of the fruit is sweet and edible, also used in flavouring.

31. WHITE DHUP, BLACK DAMMAR
Canarium strictum Roxb.

(Family: Burseraceae)

Syn. *C. resiniferum* Brace ex King

Hindi *Kala damar*; Assamese *Dhuna*; Kannada *Dhup*; Tamil *Kundrikam*; Malayalam *Thelli*.

Field Identification: Lofty, resinous, buttressed tree up to 40 m tall, young foliage crimson, rusty hairy. Mature leaves 30-60 cm, leaflets 3-7 pairs with an odd terminal. Flowering panicles 20-30 cm, Flowers brownish yellow. Petals 0.6 x 0.2 cm oblong. Fruit stony, egg-shaped, 3.75 cm long, tapering at both ends, stone hard and bony. The bark of the trunk is hard and when struck with the back of an axe gives a ringing metallic sound. **Description:** Lofty, handsome, resinous deciduous tree up to 40 m high, 2.5 m in girth, buttressed at the base and with a dense umbrella shaped crown. Bark grey or blackish-brown, peeling off in rectangular flakes. Young foliage, brilliant crimson, rusty hairy beneath. Mature leaves dark green and leathery, shining, 30-60 x 4.5-9 cm, **leaflets** 3-7 pairs and an odd terminal one, ovate-oblong, opposite or alternate, finely round-toothed, with 10-15 pairs of lateral nerves, stalks of leaflets 0.3-0.8 cm long. **Flowering** panicles 20-30 cm long, flowers brownish-yellow, calyx cup-shaped, 2-3 mm, petals oblong twice or thrice as along as the calyx. Female flower 1.25 cm long. Male flower 0.8 cm long. **Fruit** stony, ellipsoid or egg-shaped, 3.7-5 cm long, tapering at both ends, stone hard and bony. It is a fast growing tree, attaining a height of 1.8-2.7 m in the first year. **Distribution:** Evergreen forests of the Western Ghats from Konkan southwards to Kerala, NE. India mostly in Arunachal Pradesh, Assam and Khasi Hills, along river banks and margins of evergreen forests. **Phenology: Flowers** February-April. **Fruit** ripens in the cold season. **Miscellaneous:** The trunk exudes a blackish resin which has a pleasant smell and when burnt is used as incense or to drive mosquitoes away. It is sold locally in lumps for use as incense and for making torches. The wood makes the strongest plywood boxes because of its glue-holding quality.

Fig. 12

S.No. 32. *Boswellia serrata*. 1) Leaf and flowering shoot x 1. 2) Flower x 3.

32. SALAI, INCENSE TREE, INDIAN FRANKINCENSE TREE
Boswellia serrata Roxb.

(Family: Burseraceae)

Hindi, Bengali, Marathi *Salai*; Tamil, Telugu *Parangisambrani*

Fig. 12

Field Identification: Deciduous tree up to 9-15 m. Bark thin, greenish ash-coloured, peeling off in thin flakes exuding pinkish drops of resin when cut. Leaves 20-45 cm, at ends of branches, leaflets 17-31, 2.5-8 x 0.8-1.5 cm, lance-shaped with a round toothed margin. Flowers white, sepals and petals 5-7, the latter 0.5 cm long, free. Fruit 1.3 cm long, three angled with 3 valves and 3 heart-shaped, 1-seeded nutlets winged along the margins. **Description:** A moderate-sized, deciduous tree up to 9-15 m with a spreading crown. **Bark** thin, greenish ash-coloured, peeling off in thin smooth flakes, exuding pinkish drops of resin when cut. **Leaves** alternate, 20-45 cm long, crowded at the end of branches, leaflets 17-31, opposite or nearly opposite with an odd terminal one, stalkless, 2.5-8 x 0.8-1.5 cm, lance-shaped, with a round toothed margin. **Flowers** white in 10-20 cm long panicles, calyx 5-7, petals 5-7, 0.5 cm long. Stamens 10-12, inserted at the base of a red fleshy disk. **Fruit** a 3-valved capsule, 1.3 cm long. **Seeds** 3, enclosed in 3 heart-shaped, bony and winged stones. **Distribution:** Common in moist forests of Madhya Pradesh, less in Uttar Pradesh, foothills of Himachal Pradesh, North Gujarat, Maharashtra, and South India. **Phenology:** The leaves turn yellowish to light brown in December. Leafless from January to April, new leaves appear in May-June. **Flowers** February-April. Fruit ripens May-June. **Miscellaneous:** Porcupines damage trees by gnawing on the bark. Pigs dig out seedlings. It grows readily from cuttings and coppices as well. Used for match splints and boxes, sheaths of knives and swords, a valuable material for paper pulp. The tree yields a yellowish-green oleoresin with an agreeable scent when burnt. It is a good substitute for the imported Canada Balsam.

33. NEEM, MARGOSA TREE
Azadirachta indica A. Juss.

(Family: Meliaceae)

Hindi, Bengali *Neem*; Gujarati *Limdo*; Tamil, Telugu, Malayalam *Vepu maram*; Burmese *Tama*.

Fig. 13

Field Identification: Tree with feathery foliage 20-37 cm long. Leaflets toothed in 4-7 pairs, 2-8 x 1-3 cm with a terminal leaflet. Flowers small, 0.5 cm, fragrant,

FIG. 13

S.No. 33. *Azadirachta indica*. 1) Flowering twig x $\frac{3}{4}$. 2) Flower x 3. 3) Fruits x $\frac{3}{4}$.

cream-coloured with a staminal tube. Fruit 1-2 cm long, egg-shaped, yellow when ripe with a hard stone enclosing 1-2 seeds. Handsome when in bloom with fragrant flowers and shining foliage. **Description:** A moderate-sized, fast growing tree forming a dense crown. **Bark** bitter, fissured, dark grey outside and reddish brown inside. **Leaves** 20-37 cm long, crowded towards the ends of branches, leaflets 4-7 pairs, 2-8 x 1-3 cm generally with a terminal leaflet; leaflets sub-opposite, sickle-shaped, with a toothed margin, pointed, bright green and shining, stalks of leaflets short. **Flowers** cream-coloured, fragrant, honeyed 0.5 cm long, in axillary bunches. Sepals and petals 5, the latter spoon-shaped, oblong. Staminal tube dentate. **Fruit** fleshy, 1-2 cm long, egg-shaped, turning yellow on ripening with an inner hard stone enclosing usually one and sometimes two seeds. **Distribution:** Wild in the dry forests of the Deccan and Karnataka and in the dry forests of Myanmar and Sri Lanka. It is now found wild and cultivated throughout India in areas free of frost. **Phenology:** Leafless in dry localities for a brief period. New **leaves** appear in March-April. **Flowers** March-May. **Fruit** ripens June-August. **Miscellaneous:** The ripe fruits are relished by birds and hence get dispersed widely. Seeds are eaten by rodents. The foliage is browsed by camels, nilgai and sambar; chital and hares nibble the saplings. It is an excellent avenue tree. The leaves repel insects and its mere presence is believed to keep an area free from malaria. The twigs are used as chew sticks or indigenous tooth brushes. The wood is used in carving, cigar boxes and cupboards as it is insect repellent. **Etymology**: Derived from Persian '*azad darakht*' meaning 'noble tree'. Now much sought after internationally for alternative therapies in AIDS and leprosy. Seeds yield margosa oil which is used in soap manufacture and has medicinal properties; principal insecticidal component is azadirachtin.

34. TOON, RED CEDAR
Toona ciliata M. Roem.

(Family: Meliaceae)

Syn. *Cedrela toona* Roxb.

Hindi *Toon*; Marathi *Todu*; Kannada *Tundu*; Tamil *Thevatharam*; Malayalam *Mathagiri*; Burmese *Tawtama*.

Fig. 14

Field Identification: Deciduous tree 20-30 m, 1.8-3 m in girth. Bark dark grey, blaze pinkish, fibrous, bitter to taste, juice turning purple. Leaves 30-50 cm, leaflets 11-28, 5-15 x 2-6 cm, lance-shaped, pointed. Flowers small, creamish, honey scented, petals 0.5 cm, stamens 5, free. Fruits brown capsular, 1.8-2.5 x 0.5-0.8

Fig. 14

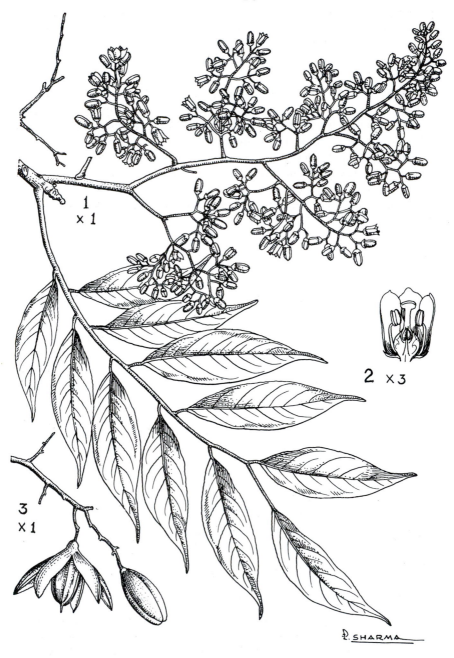

S.No. 34. *Toona ciliata*. 1) A shoot, showing flowers and leaves x 1. 2) Flower in sections x 3. 3) Fruits x 1.

cm. Seeds pale brown, winged 1.3-1.5 cm long. **Description:** A large deciduous tree 20-30 m high and 1.8-3 m in girth. **Bark** dark grey or reddish brown, exfoliating in woody scales. Blaze fibrous, pinkish with white bands on the outside, turning brown on exposure, bitter to taste. Juice turns purple on the knife blade. **Leaves** 30-50 cm long, leaflets 11-29 with a terminal leaflet, alternate or opposite, 5-15 cm x 2-6 cm, lance-shaped, pointed, margin smooth or wavy, stalk of leaflet 0.3-1.3 cm. **Flowers** small, honey scented, creamish in sub-erect or drooping terminal panicles, calyx divided to the base, petals 0.5 cm long, egg-shaped. Stamens free. **Fruit** a capsule, brown 1.8-2.5 x 0.5-0.8 cm. **Seeds** pale brown, light, winged on both ends, 1.3-1.5 cm long, including the wing. **Distribution:** All along the Sub-Himalaya, Assam, Khasi Hills, Manipur, Bihar, Madhya Pradesh, Western and Eastern Ghats, Nepal, Bangladesh and Myanmar from sea level to 1,200 m. **Phenology:** In northern India leaves commence falling in December. Leafless in January-February. **Flowers** February-April. **Fruit** ripens April-July. The winged seeds escape at different times and being light are carried far and wide by the wind. In Central and Southern India **Flowers** November-May. **Fruits** March-October. **Miscellaneous:** It is a popular avenue tree in North India. The spectacular fox-tailed orchid *Rhynchostylis retusa* grows on the trunks of 'Toon' in Dehra Dun and elsewhere. Used for furniture, cigar boxes, tea-chests, etc.

35. INDIAN JUJUBE
Ziziphus mauritiana Lamk.

(Family: Rhamnaceae)

Hindi, Punjabi *Ber*; Gujarati, Marathi *Bor*; Bengali *Boroi*; Tamil *Elandai;* Telugu *Keegu*.

Fig. 15

Field identification: Spiny shrub to a small tree with drooping branches. Bark grey, furrowed, leaves dark green above, white or buff hairy beneath, alternate, 1.8-6 x 2.5-7.5 cm, egg-shaped or half circular. Flowers greenish yellow, petals alternating with larger sepals, stamens opposite to hooded petals. Drupe 1.3-2.5 cm, round or egg-shaped, reddish brown with a single stone surrounded by edible pulp. **Description:** A shrub or small tree 6-8 m with a short bole and a spreading crown of dark green foliage with drooping branches armed with prickles. **Bark** grey, brown, or blackish and furrowed. **Leaves** dark green, alternate, variable from egg shaped-oblong to half-circular 1.8-6 x 2.5-7 cm, margin smooth or round toothed, densely hairy beneath. **Flowers** greenish yellow, sepals 5, alternating with the 5 smaller hooded petals. Stamens 5, opposite and enclosed by the petals. Disk fleshy, 10-lobed. **Drupe** 1.3-2.5 cm long, reddish brown on ripening,

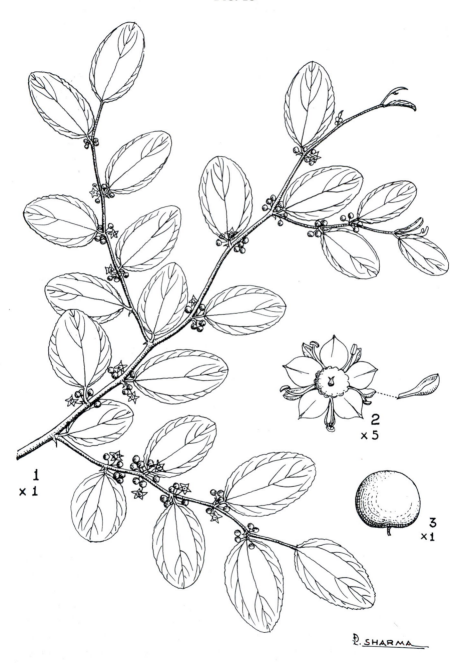

Fig. 15

S.No. 35. *Ziziphus mauritiana*. 1) Flowering shoot x 1. 2) Flower x 5. 3) Fruit x 1.

round or egg-shaped with a single stone surrounded by fleshy pulp. Stone wrinkled, bony, mostly 2-celled. **Distribution:** Throughout India, Pakistan and Myanmar in the drier regions and ascending to 1,400 m in the Himalaya. **Phenology:** The old leaves fall about March-April. New leaves appear at the same time. **Flowers** April-October. **Fruit** ripens October-March. **Miscellaneous:** Fruit edible; leaves are fed to silkworms. Wood used in agricultural implements and is an excellent fuel.

36. SOUTH INDIAN SOAPNUT, SOUTH INDIAN RITHA
Sapindus emarginatus Vahl

(Family: Sapindaceae)

Tamil *Ponam-Kottai*; Telugu *Kun-Kudu;* Gujarati *Aritha.*

Fig. 16

Field Identification: Tree with a spreading crown. Leaves alternate, leaflets 2-3 pairs, sub-opposite, terminal pair longer 7.5-17.5 cm, elliptic, sometimes notched at apex. Flowers white, 0.4-0.5 cm, in rusty hairy clusters. Petals 4-5, clawed, usually with 2 hairy scales. Drupes 2 or 3, fleshy, slightly united, 1.3-2 cm across. Seeds blackish, smooth, hard. **Description:** A medium to large tree, with a spreading crown. **Bark** rough, grey, peeling off in oblong plates. **Leaves** alternate, 12-30 cm long, leaflets sub-opposite, 2-3 pairs, terminal pair longest, 7.5-17.5 cm long, elliptic, sometimes notched at the apex; stalks of leaflets 0.3 cm long. **Flowers** white, 0.4-0.5 cm long in terminal rusty hairy clusters, males numerous, bisexual flowers few, both kinds are found in the same clusters. Sepals 5, hairy outside, 0.4 cm long, petals 4-5, shortly clawed, narrower than the sepals, usually with 2 hairy scales attached to each side of the petal. **Drupes** 2 or 3 fleshy, slightly united, 1.3-2 cm in diameter. **Seeds** blackish, smooth, about the size of a large pea, very hard. **Distribution:** South India, at low elevations in the dry deciduous forests of the Deccan, Karnataka, Nilgiris, Eastern Ghats, Sri Lanka and Myanmar. **Phenology: Flowers** October-December. **Fruit** ripens February-May, having a scent of overripe strawberries. **Miscellaneous:** Fruit is a substitute for soap for washing silk and woollens and used as shampoo by women. In North India we find an allied tree, *S. mukorossi* the soapnut tree of North India. **Etymology:** *Sapindus* refers to soap and *emarginatus* is derived from the notched apex of the leaflets.

Fig. 16

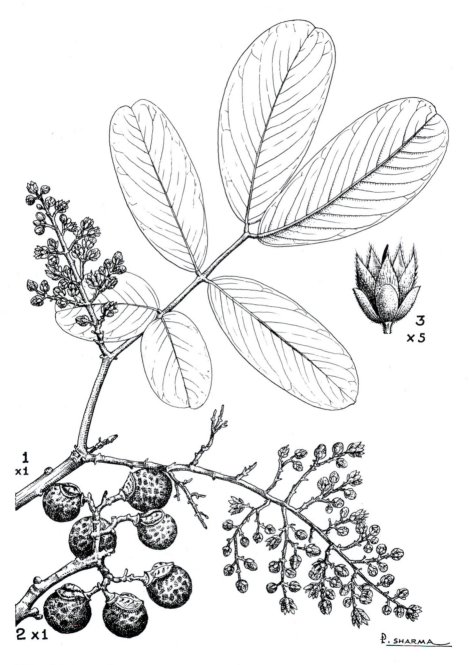

S.No. 36. ***Sapindus emarginatus***. 1) A shoot showing flowers and a leaf x 1. 2) Fruits x 1. 3) Flower x 5. (From *J S Gamble* 10971 and *K N Subramanian* 3001).

37. KUSUM, CEYLON OAK
Schleichera oleosa (Lour.) Oken

(Family: Sapindaceae)

Syn. *S. trijuga* Willd.

Hindi, Bengali, Marathi, Gujarati *Kusum*; Tamil *Puvathipuvam*;
Telugu *Posuku*; Malayalam *Puvam*; Burmese *Gyo*.

Fig. 17

Field Identification: Tree, often fluted. Blaze fibrous, pinkish-pale orange turning brown. Leaflets 2-4 pairs, lowest smaller than terminal. Flowers minute, stalkless, yellow green in short dense clusters. Fruit ellipsoid, hard skinned, pointed at tip, 1.5-3.8 x 1-1.8 cm. Seeds 1.5 cm, smooth, brown enclosed in edible pulp with a pleasant taste. **Description:** A medium-sized to large deciduous tree (sometimes evergreen) up to 20 m in height and 2.4-3.6 m girth, often with a fluted bole and a dense, spreading crown. **Bark** smooth, grey or pale brown, reddish inside. Blaze fibrous, pinkish, stippled with pale-orange, darkening to brown. **Leaves** alternate, 20-40 cm long, leaflets 2-4 pairs, opposite, those of the lowest pair smallest, 2.5-7.5 cm long, terminal pair longest, up to 15 cm long. **Flowers** minute, yellowish green, stalkless, borne in short dense clusters. Calyx small, 4-6-lobed. Petals O. Stamens 6-8 longer than calyx. **Fruit** smooth, or slightly prickly, ellipsoid, hard-skinned berry with a pointed tip, 1.5-3.8 x 1.1-1.8 cm, dry, indehiscent, 1-2 seeded. **Seeds** up to 1.5 cm, smooth, brown, enclosed in a pulpy covering which is edible and has a pleasant taste. **Distribution:** Foothills of the NW. Himalaya up to 900 m, Central and South India, Sri Lanka and Myanmar. **Phenology:** Old leaves fall from December-February, turning pale or golden yellow, leafless for a short time. New **leaves** of various shades of red, light green and dark green appear in March-April. **Flowers:** March-April. **Fruit** ripens June-July and quickly falls to the ground. **Miscellaneous:** Agricultural implements, rice pounders. In Kolar gold mines used as side props in shafts. It is a prized host tree of the lac insect.

38. WEST HIMALAYAN HORSE CHESTNUT
Aesculus indica Colebr ex Camb.

(Family: Hippocastanaceae)

Hindi *Bankhor*; Punjabi *Khanor;* Kashm. *Kakra.*

Field Identification: Leaves digitate, leaflets 5-9, 15-25 x 1.5-5 cm, margin sharply saw-toothed. Flowers white, red and yellow tinged, on large erect panicles 25-40 cm long. Petals 4, 2 narrower. Stamens 7 or 8, filaments longer than petals. Fruit

FIG. 17

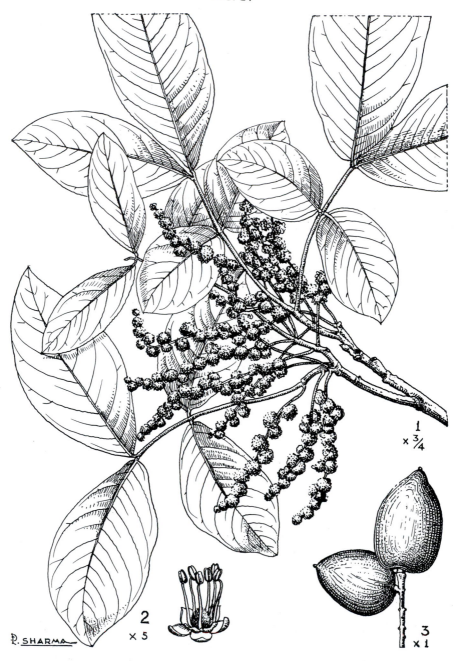

S.No. 37. **Schleichera oleosa**. 1) Flowering shoot and leaves x $\frac{3}{4}$. 2) Male flower x 5. 3) Fruits x 1.

leathery brown, 2.5-5 cm rough. Seed usually one, dark brown, shining, 3 cm across. **Description:** A medium-sized deciduous tree with scaly buds; old **bark** peeling off upwards in long thick bands. **Leaves** opposite on long stalks 10-15 cm, digitate. **Leaflets** 5-9, 15-25 x 1.5-5 cm, the centre ones the largest, oblong or oblong-lance-shaped, pointed at tip, margin sharply saw-toothed, stalks of leaflets 0.5-2 5 cm long; bud scales 4 cm long, membranous, falling off. **Flowers** white, asymmetrical, in large terminal narrow-pyramidal clusters, 25-30 cm long, 5-lobed. Calyx tubular; petals 4, white, streaked with red and yellow, 1.5-2 cm long, 2 narrower. Sta᠎ ns 7 or 8, filaments longer than petals, curved upward. Disk one-sided. **Fruit** leathery, 2.5-5 cm long, irregularly egg-shaped, brown. **Seed** usually one, dark brown, shining, 3 cm across. **Distribution:** NW. Himalaya, from Pakistan to Central Nepal at 1800-3000 m in moist shady ravines. **Phenology:** Leaves are shed in October. New leaves in April. **Flowers** April-June. Fruits July-October. **Miscellaneous:** Wood is turned into cups and dishes. Fruits and foliage used as fodder and the embryo when mixed with flour is eaten by hill people. A good avenue tree for the hills.

39. EAST HIMALAYAN HORSE CHESTNUT
Aesculus assamica Griff.

(Family: Hippocastanaceae)

Syn. *A. punduana* Wall. ex Hiern.

Bengali, Nepali *Satpate*; Burmese *Yemyaw*.

Field Identification: Deciduous tree with a large hemispherical crown. Leaves long, stalked, up to 28.5 cm, palmate, digitate, leaflets 6, 30 x 11 cm, pointed, margins lightly toothed, saw-like. Flowers creamish tinged with brown-red, in large erect terminal panicles up to 58 cm long. Fruit large, leathery, 5-9 cm, rusty brown. **Description:** A handsome deciduous tree with a wide spreading hemispherical crown 12-15 m tall. **Bark** greyish, when chipped it reveals an orange and white blaze with a yellow ring. **Leaves** opposite, palmately compound, digitate, thinly leathery, leaflets 6, 30 x 11 cm broad round and pointed at the tip, margin thinly saw-toothed, almost stalkless to 1.2 cm long. **Flowers** showy, borne on very large terminal erect panicles up to 58 cm. Petals 5, often 4, cream coloured tinged with brown-red. Stamens 8, filaments long, 3.5 cm, curved like a horse shoe. **Fruit** large, leathery, 5-9 cm, rusty brown. **Seed** 2.5 cm across, shining. **Distribution:** E. Himalaya, North Bengal, Sikkim, Arunachal Pradesh and Khasi Hills, Bhutan, Upper Myanmar. Mostly near streams up to 1,000 m. **Phenology:** **Leaves** turn yellow in autumn and are shed in November. New leaves appear in February. **Flowers** February-March to early April. **Fruit** ripens August-October.

Miscellaneous: Handsome tree, worthy of cultivation in parks. It comes up well in North India from 500-1000 m in areas of good rainfall and does extremely well in Dehra Dun, producing fertile seeds. Raised from seed.

40. MAPLE
Acer caesium Wall. ex Brandis

(Family: Aceraceae)

Hindi *Kainsal*; Punjabi *Trekhan*.

Fig. 18

Field Identification: Tree with 5 unequally lobed leaves 8-15 cm long, pointed, broader than long, with a heart-shaped base and saw-toothed margin; leaf stalk red. Flowers greenish, 5 mm across in flat-topped clusters. Fruit with 2 wings, wings 3.8-5 cm long, nuts dark brown with a hump-like swelling. **Description:** A tall handsome, deciduous tree up to 25 m, trunk often covered with burr-like growth. **Bark** thin grey, peeling off in thin small strips. **Leaves** opposite, large 8-15 cm long, broader than long, with a heart-shaped base, and 5 unequal lobes, saw-toothed on the margin, usually pale and with greyish bloom beneath, red, when young, petioles red. **Flowers** greenish yellow, 5 mm across, appearing with young leaves, in flat-topped clusters. **Fruit** with 2 divergent dragonfly-like wings 3.8-5 cm long, nuts dark brown, with a hump-like swelling. **Distribution:** Kashmir eastwards. Pakistan, Central Nepal, at 2,200-3,000 m in fir forests on open grassy places, more or less gregarious. **Phenology:** Leafless in winter, leaves are shed in November. New leaves red and appear from March-May. **Flowers** March-May. **Fruit** ripens from July to October. **Miscellaneous:** Carvings, for making bowls, furniture, butts of rifles. Makes excellent moisture-proof plywood.

41. CASHEWNUT
Anacardium occidentale L.

(Family: Anacardiaceae)

Hindi, Gujarati, Marathi *Kaju*; Bengali *Hijli Badam*; Malayalam *Parangi Mavu*; Tamil *Munthiriparuppu*.

Fig. 19

Field identification: Small tree with thick leathery leaves, 7-18 x 4.5-10 cm, notched at the apex. Flowers yellow with pink stripes, 1-2 cm wide. Nut kidney-

FIG. 18

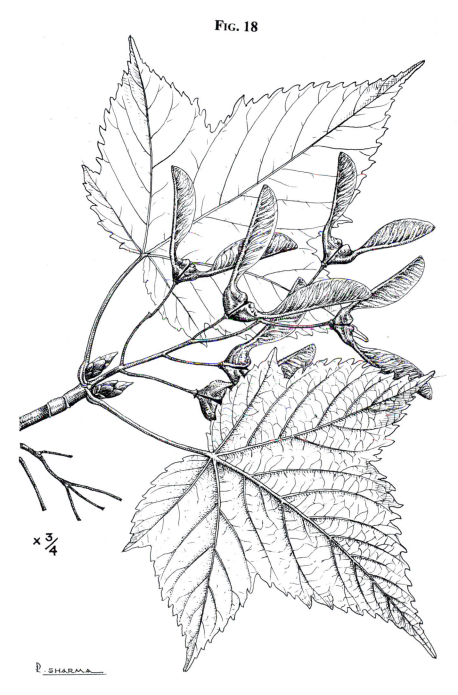

S.No. 40. *Acer caesium*. Leaves, fruits with dragonfly-like wings for gliding x $\frac{3}{4}$. (From *K.C. Sahni* 26837).

FIG. 19

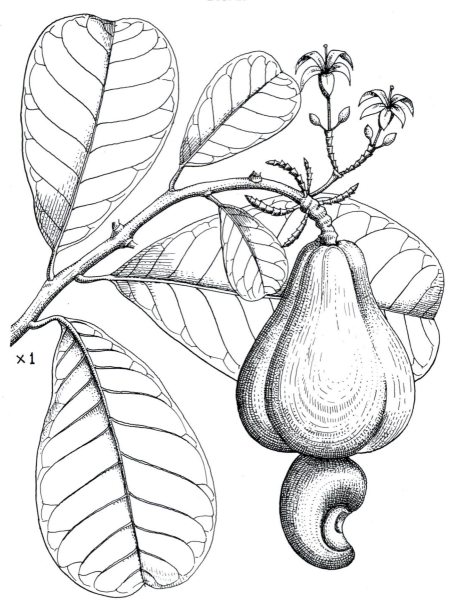

×1

P. SHARMA

S.No. 41. *Anacardium occidentale*. Fruit and nut x 1, with a branch sprouting new leaves behind.

shaped, 2-3.8 cm seated on flower stalk much enlarged into a fleshy orange-red 'apple' 5-8 cm long. **Description:** A small evergreen tree with smooth brown bark. **Leaves** leathery, broad and often notched at apex, 7-18 x 4.5-10 cm, leaf stalk 1.2-1.8 cm. **Flowers** yellow, borne on 23 cm long terminal panicles with a heavy sickly sweet smell. Flowers 1-2 cm wide, sepals 5, 0.4-0.5 cm, petals 5, linear 0.8-1.2 cm, yellow with pink stripes. Stamens 7, 1 cm long. The nut is kidney-shaped, 2-3.8 cm, seated on the flower stalk and enlarged into a fleshy orange-red 'apple' 5-8 cm long. **Distribution:** Introduced by the Portuguese from Brazil centuries ago in Goa and now common in coastal areas in Southern India, Central India and Orissa. **Phenology: Flowers** December-April. **Fruit** ripens March-June. **Miscellaneous:** Direct sowing of seeds at the commencement of the monsoon. The seeds are planted below the soil. **Uses:** The kidney-shaped nut is the commercial cashew nut 'kaju' which is edible and much prized all over the world. The fleshy-orange red 'apple' is juicy, rich in vitamin C and alcoholic, and makes a beverage called 'feni' which is popular in Goa. It is also a useful tree for reclamation of sand dunes in coastal areas.

42. JHINGAN, WODIER WOOD
Lannea coromandelica (Houtt.) Merr.

(Family: Anacardiaceae)

Syn. *Lannea grandis* (Densl.) Engler

Hindi *Jhingan*; Marathi *Moi*; Bengali *Jiyal*; Telugu *Appriyada*; Tamil *Wodier*; Malayalam *Odiya Maram;* Kannada *Ajasringi.*

Fig. 20

Field identification: Small to medium sized tree with grey bark, red inside. Leaves with 5-11, pointed, egg-shaped leaflets 6.3-15 x 2.5-8.8 cm. Flowers 0.4-0.5 cm wide, yellow-green, male and female on different branches. Sepals and petals 4 each, stamens twice the number of petals. Drupes 1.3-1.5 cm, with dull red pitted stone. **Description:** A small to moderate sized deciduous tree. **Bark** grey, peeling off in thin round plates, red inside. **Leaves** up to 45 cm long, leaflets 5-11, 6.3-15 x 2.5-8.8 cm, egg-shaped, pointed at the tip, short stalked, 0.3 cm long. **Flowers** 0.4-0.5 cm, unisexual, yellowish green, male and female on different branches. Calyx 4-lobed, petals 4, greenish yellow. In male flowers, stamens twice the number of petals. Female flowers, ovary stalkless, with 3-4 distinct styles. **Drupe** 1.3-1.5 cm, smooth, dull red with a pitted stone. Handsome when covered with feathery cream coloured blossoms. **Distribution:** Sub-Himalaya and throughout the hotter parts of India, extending southwards to Kerala. **Phenology:** Leafless from

Fig. 20

S.No. 42. *Lannea coromandelica* 1) A twig bearing female flowers and fruit x 1.
2) Male flower x 5. 3) Female flower x 5. (From *Lace* 932 and *Campbell* 7529).

November-May, turning yellow before falling. New leaves in May-June. **Flowers** February-April. **Fruit** ripens May-July. **Miscellaneous:** Carving, turnery and furniture. The tender shoots are relished by elephants and the leaves are greedily eaten by cattle. Jhingan gum tapped from the trunk is used in calico printing. Crows are fond of the fruit; they swallow the pulp and help in the dispersal of the seed.

43. MANGO TREE
Mangifera indica L.

(Family: Anacardiaceae)

Punjabi, Hindi, Bengali *Aam*; Gujarati, Marathi *Amba*; Tamil, Malayalam *Mamaram*; Telugu *Mamidi.*

Fig. 21

Field Characters: Large tree with a dome-shaped crown. Easy to recognize from other members of the mango family in evergreen forests by the presence of only one stamen in small yellow-greenish flowers which are borne on cone-shaped panicles. Fruit 5-20 cm, fleshy, yellow when ripe. **Description:** A large evergreen tree, up to 45 m high with a heavy dome-shaped crown, often with a girth of 3.6 m and over. **Bark** rough, thick, dark grey and fibrous. **Leaves** crowded at the ends of branches 10-30 x 2-10 cm, oblong or lance-shaped, pointed at the tip, dark glossy green, pinkish when young, with an aromatic resinous odour when crushed, leaf stalk 2.5-6 cm long, swollen at the base, young leaves hang vertically downwards, while the colour is pink. Flowering shoots conical, **flowers** 0.4 cm across, greenish yellow, scented, male and bisexual on the same panicle, calyx 4-5 falling off, petals 4-5, pale yellow, longer than sepals, disk fleshy, stamens 4-5, only one stamen is fertile and much longer than the rest. **Drupe** 5-20 cm, fleshy, yellow when ripe, stone compressed fibrous. Many varieties are found in different parts of the Subcontinent and they vary in flavour and quality. **Distribution:** In cultivation in the Subcontinent, for at least 4,000 years. It probably originated in the Assam-Arunachal-Myanmar region where wild trees of *M. indica* and *M. sylvatica* have been recorded. Wild in the foothills of the Eastern Himalaya and the Western Ghats. It has run wild throughout the Subcontinent in tropical and subtropical areas. **Phenology:** Evergreen. **Flowers** January-March. **Fruit** ripens April-July. **Miscellaneous:** It is one of the best fruits of the world. It is a popular avenue tree on national highways as it gives dense shade. In May, the branches and trunks of mango are often covered with the spectacular fox-tailed orchid in Dehra Dun. A tree in Chandigarh has a trunk of 9.6 m girth and a crown-spread of 2,250 sq. m with an annual yield of 17 metric tonnes of fruit.

Fig. 21

S.No. 43. **Mangifera indica**. 1) Leaves and inflorescence x 1. 2) Flower x 5.
3) Open flower characterised by only stamen x 5.

44. MARKING-NUT TREE
Semecarpus anacardium Linn. f.

(Family: Anacardiaceae)

Hindi, Gujarati, Marathi *Bilwa*; Bengali *Bela*; Tamil *Shenkotta*;
Malayalam *Chera*; Burmese *Thitsi-bo*.

Fig. 22

Field Identification: Bark with acrid juice. Blaze red. Fruit 2.5-3.8 cm shining purplish black seated on a fleshy coloured stalk. **Description:** A medium sized deciduous tree with a low spreading crown. **Bark** dark brown with acrid juice. Blaze red, exuding resin which blackens on exposure. **Leaves** large and leathery, clustered near the ends of branches, 15-45 x 10-20 cm, oblong, broad and round towards the apex, base wedge-shaped, lateral nerves 15-25 pairs, petiole thick 1.25-2.5 cm long. Underside of leaves and flowering shoot rough, hairy. **Flowers** 0.5-0.8 cm across, greenish yellow, sessile in terminal panicles, calyx 5-cleft, petals 5, much larger than sepals. Stamens 5, inserted on the broad collar-like disk. **Fruit** 2.5-3.8 cm long, shining, purplish black, seated on a fleshy orange-coloured stalk which is edible. **Distribution:** Sub-Himalayan tracts eastwards to Khasi Hills, Bangladesh and Myanmar. Central India down to Kerala. **Phenology:** Old leaves turn yellow before falling, shed in February-March. New leaves appear in May. **Flowers** May-August. **Fruit** ripens December-March. **Miscellaneous:** The black resin from the fruit is used by Indian washermen as marking-ink for clothes. The orange-coloured receptacle is eaten either dry or roasted.

45. DRUMSTICK TREE
Moringa oleifera Lamk.

(Family: Moringaceae)

Syn. *M. pterygosperma* Gaertn.

Hindi *Sainjna*; Bengali *Sajina*; Malayalam *Muringa*; Tamil *Murunkai*.

Fig. 23

Field Identification: Small to medium-sized deciduous tree with corky bark. Leaves alternate, compound, leathery, leaf stalk swollen at base. Flowers creamish, 2.5 cm across, in clusters. Fruit long, 22.50 x 1.5-2 cm, green, hanging downward, ribbed slightly, constricted between seeds, seeds winged, 2.5 cm

Fig. 22

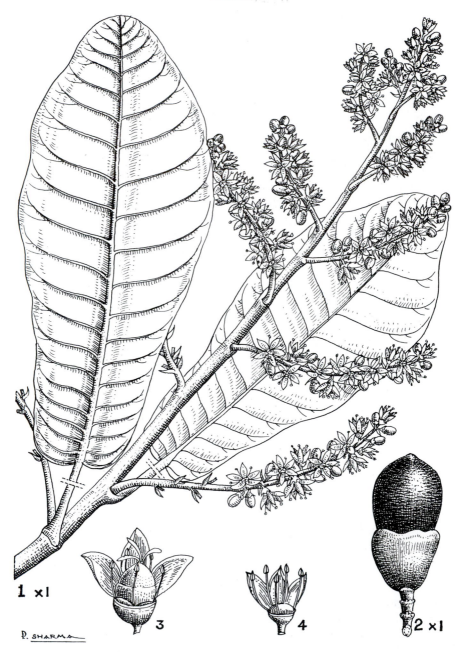

P. SHARMA

S.No. 44. ***Semecarpus anacardium***. 1) Leaves, inflorescence x 1. 2) Fruit x 1. 3) Portion of hermaphrodite flower, enlarged. 4) Portion of male flower with stamens and petals.

FIG. 23

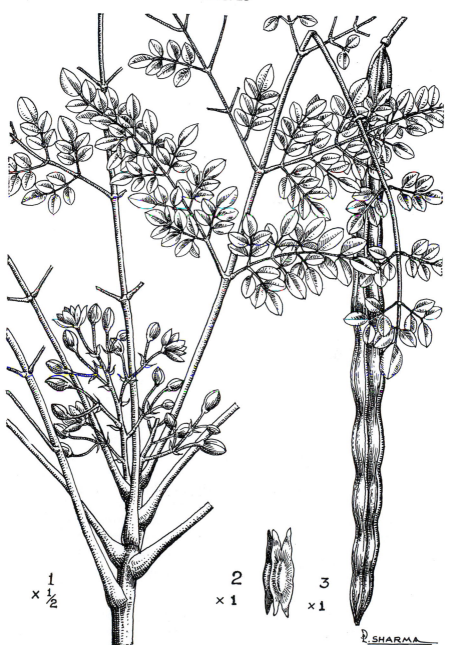

$\times \frac{1}{1}$ $\frac{1}{2}$

2 $\times 1$

3 $\times 1$

R. SHARMA

S.No. 45. *Moringa oleifera*. 1) Leaves and inflorescence x $\frac{1}{2}$. 2) 3-cornered winged seed x 1. 3) Fruit x 1.

long. **Description:** A small or sometimes medium-sized tree up to 15 m tall. **Bark** very pale brown, corky and deeply fissured, blaze crisp, pale yellow with a few narrow dark purple bands and mottled with orange yellow. **Roots** pungent. **Leaves** alternate, 30-90 cm, usually thrice compound; **Leaflets** feathery, elliptic 1-2.5 cm, opposite, terminal larger. **Flowers** 2.5 cm across, honey scented. Sepals linear-lance-shaped, bent. Petals 5, 1.8-2.3 cm, linear spoon-shaped, white with yellow dots. **Fruit** 20-50 cm long by 1.5-2 cm wide, capsule 9-ribbed, constricted between the seeds, greenish. **Seeds** 2.5 cm long, winged, 3-cornered. **Distribution:** Sub-Himalaya from Jammu and Kashmir to Bengal, Kutch, Gujarat, Pakistan. Cultivated elsewhere. **Phenology: Leaves** turn yellow and fall in December-January. New leaves in February-March. **Flowers** January-March. **Fruit** ripens April-January. **Miscellaneous:** By direct sowing of seeds. Cuttings of large size up to 3 m strike roots readily and grow to sizable trees in a few months. The leaves are relished by camels. The branches are broken off by sambar, deer and monkeys. Porcupines and pigs feed on the soft roots. The flowers and pods are esteemed as a vegetable. The oil from the seed is edible and is also used in cosmetics.

46. BABUL

Acacia nilotica (L.) Del. ssp. *indica* (Benth.) Brenan

(Family: Mimosoideae)

Hindi, Bengali, Marathi *Babul*; Gujarati *Baval*; Punjabi *Kikar*; Tamil *Karuveli*; Telugu *Nallatumma*.

Fig. 24

Field Identification: The pods are diagnostic and help in the recognition of Babul from other Acacias. Moderate sized tree with dark brown vertically fissured bark. Sharp pointed, white, paired spines 5 cm long, feathery bipinnate leaves, yellow sweet scented flowers in globose heads 1.3 cm across, and necklace-like pods 7.5-15 x 1.25 cm, constricted between seeds. **Description:** A moderate sized tree with a short thick cylindrical trunk. **Bark** dark-brown or nearly black, with vertical fissures which sometimes run spirally up the tree. Characterized by white sharp-pointed, paired spines, up to 5 cm long, feathery bipinnate **leaves,** leaf stalk 2.5-5 cm long, pinnae 3-10 pairs, 3.8 cm long. **Leaflets** 10-20 pairs, 2 to 5 mm long, linear. **Flowers** golden yellow, sweet scented, in globe-like heads 1.3 cm across, flower stalk 1.3-2.5 cm long. Corolla yellow, bell-shaped, twice as long as calyx, 2.5-3.5 mm long. **Pod** densely grey-downy, 7.5-15 x 1.25 cm, necklace-shaped, constricted between the seeds. **Seeds** 8-12 per pod, compressed, ovoid, 7-9 x 6-7 mm, dark brown, shining. **Distribution:** Indigenous to Sind where they occupy the largest tracts. Rajasthan, Gujarat and the Deccan; naturalised

FIG. 24

S.No. 46. *Acacia nilotica* ssp. *indica.* 1) Spines, leaves and globose flowers, with pod (5) containing 8-12 seeds x 1. 2) Flower bud. 3) Corolla. 4) Pistil all enlarged.

in the rest of India, Sri Lanka. **Phenology:** New leaves appear in March-April. **Flowers** June to September. **Fruit** ripens in December-January. **Miscellaneous:** The wood is durable and resistant to termites. It is used as railway sleepers and is a high quality fuel. Bark and pods are esteemed as tanning materials. Leaves and pods are good fodder. The spines serve as fishing hooks and paper pins.

47. KHEJRI TREE
Prosopis cineraria Druce

(Family: Mimosoideae)

Syn. *P. spicigera* Linn.

Hindi, Punjabi *Jand*; Sind *Kandi*; Rajasthan *Khejri*; Gujarati *Summi*; Marathi *Saundar*; Telugu *Jambi*; Tamil *Jambu*.

Fig. 25

Field Identification: Thorny tree with deeply fissured grey bark. Prickles broad-based, conical. Leaves two pinnate, pinnae 1-2 pairs, 2.5-7.6 cm, leaflets 7-12 pairs, 5-12 x 1-3 mm. Flowering spikes 5-13 cm. Flowers creamy white, petals 5, yellow, 2-4 mm. Pods 12-25 cm x 5-8 mm, hanging down, slender, with dry sweetish pulp, seeds 10-15. **Description:** A medium sized thorny tree with a very long tap-root. **Bark** grey, rough with deep fissures. Branches and branchlets armed with scattered broad-based conical prickles. **Leaves** alternate, bluish green, bipinnate, pinnae and leaflets opposite, pinnae 1-2 pairs 2.5-7.6 cm long, leaflets 7-12 pairs, 5-12 x 1-3 mm, apex usually mucronate. Spikes 5-1.3 cm in short axillary panicles. **Flowers** creamy white, petals 5, yellow, 2-4 mm long, tips recurved. Stamens 10, anthers gland-tipped. **Pods** 12-25 cm x 5-8 mm, leathery, slender, hanging, filled with a dry, sweetish pulp constricted between 10-15 seeds. **Distribution:** Sind, Baluchistan, Punjab, Rajasthan, Gujarat and dry parts of Central and Southern India. **Phenology:** Leafless for a short period, they begin to fall from mid-October to mid-February. New **leaves** appear in January-February **Flowers** mid-February to March. **Fruit** ripens April-May. **Miscellaneous:** One of the most useful trees of the desert. Wood used in house building, agricultural implements and is an excellent fuel. The foliage and pods are good fodder for camels and goats. Khejri is regarded as a saviour in the desert and is sacred in Rajasthan.

FIG. 25

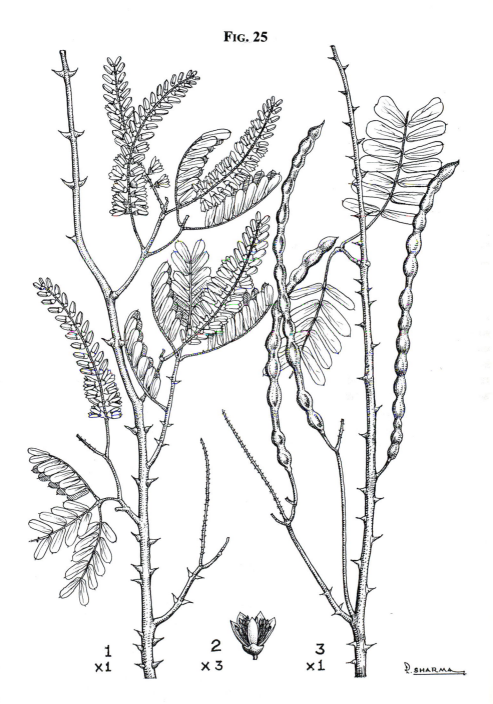

S.No. 47. ***Prosopis cineraria***. 1) Leaves and inflorescence x 1. 2) Flowers x 3. 3) Twig with pod x 1. (From *Hole* 30944).

48. KOKKO, BLACK SIRIS
Albizia lebbeck (L.) Benth.

(Family: Mimosoideae)

Hindi, Bengali, Marathi, Gujarati *Siris;* Tamil *Vagai*; Malayalam
Dirasana; Burmese *Kokko.*

Fig. 26

Field Identification: Deciduous tree. Bark dark grey or brownish. Leaves bipinnate,
pinnae 2-3 pairs, 5-20 cm, leaflets 6-9 pairs, 2.5-5 x 1.5-2.5 cm. Flowers white,
fragrant, in large globose heads, 2-4 stamens with long filaments 2.5 cm long. Pods
large, straw coloured 10-30 x 2.5-5 cm, 6-12 seeded, remain long on the tree and
rattle in the breeze. **Description:** A large deciduous tree up to 18 m high and 1.5-1.8
m in girth. **Bark** dark grey or brown. **Leaves** bipinnate, rachis 8-23 cm with a large
gland in the middle of the leaf stalk. Pinnae 2-3 pairs (rarely 4-6 pairs), 5-20 cm
long, leaflets 6-9 pairs, 2.5-5 x 1.5-2.5 cm, obliquely oblong. **Flowers** white, fragrant
in large globose heads, borne on a 5-10 cm long stalk, flower heads 2-4, stalks of
individual flowers 3 mm long. Calyx 4.5 mm, corolla about 9 mm long, teeth
triangular, stamens with the filaments united at the base and about 2.5 cm long,
anthers minute. **Pods** large, thin straw coloured, 10-30 x 2.5-5 cm, round at both
ends or pointed at both ends, 6-12 seeded, seeds 6-9 mm, compressed, pale brown.
Pods remain on the tree for a long time and rattle in the breeze. **Distribution:** Widely
distributed throughout India, Sri Lanka and Pakistan. **Phenology:** Leaves commence
falling in October-November. New leaves appear in April. **Flowers** April-May. **Fruit**
ripens in September. **Miscellaneous:** One of the important trees of Rajasthan, useful
for afforestation and as fodder for camels and cattle. In the Andamans the timber is
valuable for ornamental work and is exported to U.S.A and U.K. The burrs are
popular for carving and turnery. The planks are used for panelling. **Etymology:**
Albizia is named after Albizzi, an Italian naturalist; *lebbeck*, after a place in Egypt
where it was much planted as an avenue tree.

49. ASHOKA
Saraca asoca (Roxb.) de Wilde

(Family: Caesalpinioideae)

Hindi, Bengali, Marathi, Gujarati *Ashoka*; Malayalam *Asokam*;
Tamil *Asogam*; Burmese *Thawka.*

Field Identification: A small evergreen tree with compound leaves up to 30 cm
long. Leaflets 4 pairs without a terminal leaflet, ovate-oblong, pointed, up to 25 x

80

FIG. 26

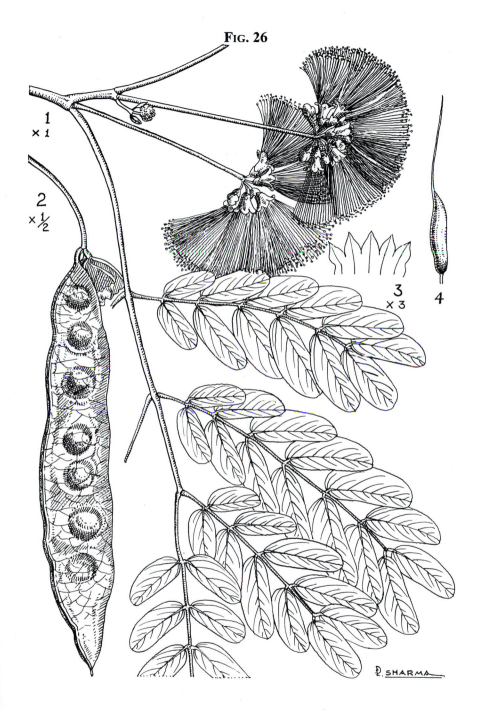

S.No. 48. *Albizia lebbeck*. 1) Flowering twig x 1. 2) Pod x $\frac{1}{2}$. 3) Corolla split open x 3. 4) Pistil (enlarged).

6.4 cm. New leaves in tassels of pink or purple. Flowers fragrant, orange yellow, in corymbs 7.5-10 cm across. Individual flowers 2.5 cm long. Calyx tube orange yellow, cylindric, petals absent, stamens 6-8 grow beyond calyx tube, anthers purple. Pods leathery, coppery-red, dark grey on maturity, 5-15 x 2.5 cm, sharply pointed at the tip, with 4 to 8 grey seeds. The pods split into two halves on ripening and coil up. **Description:** A small handsome evergreen tree 6-9 m high. Its branches form an elegant close-leaved crown. **Bark** dark brown or greyish brown. The drooping compound **leaf** is *c.* 30 cm long with 4 pairs of leaflets without a terminal one borne on a 25 cm long petiole. **Leaflets** dark green and glossy, ovate-oblong 7.5-25 x 3.5-6.4 cm with a pointed tip called a 'drip tip', an adaptation to drain off rain water. Such leaves are seen in plants of high rainfall areas. Side nerves *c.* 12 pairs, strongly netted underneath. New leaves dangle in tassels of pink or purple for several days before stiffening and straightening to maturity. The Ashoka presents like the mango, *Amherstia nobilis, Mesua ferrea* and several other evergreens, the remarkable phenomenon of pendant leaves, that is the young leaves hang down limply. The fragrant orange-yellow **flowers** appear in clusters 7.5-10 cm wide. The ashoka also produces flowers on the old wood. Individual flowers are 2.5 cm long on 1.2 cm long stalks, calyx tube cylindric, orange yellow, petals absent, stamens 6-8, projecting out, anthers purple, style curved into a ring. **Pod** coppery red, black on maturity, 5-15 x 2.5 cm, terminated by a short and sharp point, containing 4 to 8 smooth grey seeds. The pods split into two halves on ripening and coil up. **Distribution:** Wild in the evergreen forests of the Western Ghats and the Eastern Himalaya up to 750 m, also in Khasi and Garo Hills, Andamans, Myanmar and Sri Lanka. Usually in deep shade of rain forests and along streams. **Phenology:** Evergreen, new **leaves** appear in April. **Flowers** January to May. **Fruits** August-September. **Miscellaneous:** The seeds are chewed as a substitute for betel nut. In Sri Lanka the timber is used for house building and in Assam for ploughs. Planted in monastery gardens in India and Myanmar e.g. Kanheri Caves near Mumbai. It is seen sculptured on ancient Buddhist temples in Sanchi and Mathura. Buddha was born under an Ashoka tree, hence it is sacred to Buddhists. Hindus regard it as sacred, being dedicated to Kama Deva, god of love. Easily propagated from seeds. It should be planted in sheltered places to protect it from hot winds.

50. FLAME AMHERSTIA, PRIDE OF BURMA
Amherstia nobilis Wallich

(Family: Caesalpinioideae)

Burmese *Thawga.*

Field Identification: Tree with 0.45 m long pinnate leaves. Leaflets 6-8 pairs, 15-30 x 2.5-3.7 cm, tapering to a fine tip. Flowers 20-26, borne on 0.9 m drooping sprays. Flowers 20 x10 cm across, petals 5, red, 3 large unequal with 6 blobs of gold, 2 minute. Stamens 10, one short, free, and 9 united basally, 5 anthers on long filaments, 4 on stalkless anthers. Pod 12.5-17.5 cm, broad at top, crimson when young. Seeds 4-6. **Description:** A moderate sized evergreen tree 9-12 m high, rarely up to 18 m, with a round topped low crown of spreading branches and dense green foliage, much like Ashoka when not in bloom. Trunk stout with thick dark ashy-grey bark. **Leaves** exceptionally handsome, pinnate, 0.45 m long, the young leaflets hang in brownish-pink tassels. Leaflets 6-8 pairs, opposite, 15-30 x 2.5-3.7 cm, oblong, tapering to a fine 'drip tip'. The individual **flowers** 20 x 10 cm across look like humming birds, each mounted on slender intensely red stalks and arranged into 0.9 m long, drooping, candelabra-like sprays, so that the bird-like flowers stand out elegantly. The flowers have 5 red petals, 3 large with blobs of gold, reversedly heart-shaped, and the other 2 minute. The sprays support 20 to 26 beautiful flowers. The calyx is tubular at base, then divides into 4 petal-like sepals. Stamens 10, a short upper one, 9 united basally, 5 anthers on long filaments alternating with 4 almost stalkless anthers. Anthers large dark green. **Pod** 12.5-17.5 cm, broad at top with 4-6 seeds. The young pods are brilliant crimson with greenish marks. **Distribution:** Tenasserim, South Myanmar. Very rare in the wild. In 1927, R.N. Parker of the Indian Forest Service found a few trees in the Mergui District in the erstwhile Burma (Myanmar) where the locals called it 'natthami' (daughter of the spirit). **Phenology: Flowers** January-March. The flowers last 2-3 days. **Fruits** April. **Miscellaneous**: Raised from seed in pots, but seeds are difficult to obtain. It is best propagated by layering in the hot season and planting out during the rains. Besides Myanmar where it is indigenous, there are trees in cultivation in the Botanical Gardens in Calcutta, Mumbai and Sri Lanka. The seeds seldom ripen outside Myanmar. **Etymology:** The name was given in honour of Sarah, Countess of Amherst, wife of the governor of Myanmar; *nobilis,* on account of the exquisite beauty of the flowers. It is often claimed as the most beautiful flowering tree in the world. Rare and has only been sighted twice in the wild state. Wild in Yunsalin valley in Myanmar.

51. INDIAN LABURNUM, GOLDEN SHOWER
Cassia fistula Linn.
(Family: Caesalpinioideae)

Plate 6

Hindi, Punjabi, Bengali *Amaltas*; Marathi *Bahava*; Tamil *Konnei*; Telugu *Rela*; Burmese *Ngu*.

Field Identification: Clusters of drooping sprays 30-60 cm long with yellow flowers 4-6 cm across. The 30-60 cm long dark brown pods 2.5 cm wide hang like straight pipes. **Description:** A small to medium-sized tree. **Bark** smooth, pale or greenish grey. **Leaves** 23-46 cm long, leaflets 4-8 pairs, opposite, 5-12.5 x 3.7-8 cm ovate-oblong. **Flowers** 4-6 cm across, bright yellow in clusters of 30-50 cm long drooping sprays, flowers on 2.5-5 cm long stalks, calyx 8-10 mm long, divided to the base, petals 5, 1.8-2.5 cm long, obovate, shortly clawed. Stamens 10, 3 longest stamens curled, 4 smaller median and the remaining three short and erect. **Pods** 30-60 cm long, 2.5 cm wide, cylindric, smooth and dark brown when ripe. **Seeds** 40-100, immersed in dark coloured sweetish pulp and separated from one another by transverse septa. The long pods hang like straight pipes and have given the tree its Latin name *fistula*. It is a beautiful tree when in bloom, with clusters of drooping sprays of yellow flowers justifying its name of Golden Shower. **Distribution:** Found scattered in the deciduous forests of the Subcontinent including Myanmar and Sri Lanka. Up to 1,220 m in the Himalaya. **Phenology:** Leafless for a brief period between March-May. New **leaves** appear in May. **Flowers** April-July. **Fruit** ripens December-April. **Miscellaneous:** Reproduction is effected through monkeys, jackals, bears and pigs which break open the pods to eat the pulp and thus scatter the seeds. In areas frequented by monkeys many trees grow together. A popular tree for gardens and avenues. The timber is durable and is used for house posts, agricultural implements and tool handles. The pulp is an ingredient of spiced Indian tobacco. Raised from seeds sown in beds in March-April. Germination takes place early in the rains.

52. VARIEGATED BAUHINIA
Bauhinia variegata L.
(Family: Caesalpinioideae)

Hindi *Kachnar*; Nepali *Taki*; Marathi *Kanchan*; Telugu *Mandari*.

Field Identification: Trunk stocky. Leaves thick 7.5-15 cm, about as broad, cleft 1/4 to 1/3 of the way down, base heart shaped, nerves 11-15, prominent.

Flowers up to 7.5 cm across, pink, white and mauve splashed with purple. Petals 5, 3.6-6.4 cm long. Pods 15-30 x 1.8-2.5 cm, hard and flat. Seeds 10-15. **Description:** An ornamental tree with a stocky trunk and dark grey or brown bark. Leaves thick, 7.5-15 cm long and about as broad, cleft one-quarter to one-third of the way down into two obtuse lobes, base heart-shaped with 11-15 prominent nerves beneath. Flowers large, up to 7.5 cm across, fragrant, pink, white and mauve splashed with purple, calyx grey-hairy, 1.2-2.5 cm long, petals 5, 3.8-6.4 cm long. Stamens 5. Pod 15-30 x 1.8-2.5 cm, hard, flat, on a 2.5 cm long stalk. Seeds 10-15, nearly circular, 1.2-1.9 cm in diameter, flat, brown. **Distribution:** All along the Himalayan foothills, Central and South India. **Phenology:** Leafless in March. New **leaves** April-May. **Flowers** February-April, visited by bees through which pollination is effected. **Pods** ripen in May-June. They burst open on ripening and the seeds are scattered up to 5 m or more. **Miscellaneous:** Raised from seeds sown in May. A popular ornamental tree. The leaves are used to roll 'bidis' and flower buds are pickled and eaten as a vegetable. The leaves are good fodder and the timber is a good fuel. **Etymology:** *Bauhinia* is named after the botanists John and Casper Bauhin who were brothers. Almost all Bauhinias have two-lobed or twin leaves, shaped to suggest the imprint of the hoof of a camel.

53. TAMARIND
Tamarindus indica L.

(Family: Caesalpinioideae)

Hindi, Gujarati, Bengali *Imli*; Malayalam *Puli*; Marathi *Chincha*; Burmese *Magvi*.

Fig. 27

Field Identification: Tree with feathery leaves 5-15 cm, leaflets 10-20 pairs, opposite, 1-1.8 cm x 4-5 mm. Flowers in clusters of pale yellow, petals 1-1.5 cm Pod thick, curved, 7.5-20 cm, with dark brown acidic pulp traversed by fibres. Seeds 3-10, dark brown and shining, 1.3 cm across. **Description:** Handsome tree with a dome-shaped crown to 30 m high and 5 m in girth. A girth of 12.8 m is known from Sri Lanka. Bark dark grey, fissured, blaze fibrous pale red. **Leaves** thick 5-15 cm, leaflets 10-20 pairs, opposite without the terminal leaflets, 1-1.8 cm x 4-5 mm. **Flowers** in 10 cm long clusters of pale yellow blossoms. Petals 1-1.5 cm, the upper 3 developed, pale yellow, variegated with red streaks, the lower 2 reduced to scales. Perfect stamens 3, filaments united to the middle, the others reduced to bristles. **Pod** thick 7.5-20 x 2.5 cm, curved, filled with dark brown acidic pulp, traversed by fibres. **Seeds** 3-10, 1.3 cm across, hard, brown and shining.

FIG. 27

S.No. 53. *Tamarindus indica*. 1) Flowering twig x 1. 2) Stamens x 2. 3) Pod x 1.

Distribution: Native to tropical East Africa. Senegal's capital city is named after the tree, whose local name is '*Dakar*'. Common throughout the Subcontinent including Myanmar and Sri Lanka in areas free of frost. Introduced to India in very ancient times. **Phenology:** New leaves appear in March-April. When young they have a brilliant emerald hue; being feathery they give the tree a graceful appearance. **Flowers** May-June. **Fruits** March-April i.e. 10-11 months after flowering. **Miscellaneous:** Its strong, supple branches are not affected by wind and it is known as a hurricane-resistant tree suitable for avenues. A valuable timber and choice fuel, it was a major fuel for producer-gas (gasogen) units that powered Indian trucks during World War II. The pulp from the pods is used to season curries, chutneys and ice-cream. **Etymology:** The name tamarind derives from the Arabic *tamr-hindi* which means 'date of India' because its large pulp resembles preserved dates. It is, however, native to Africa and was brought to India by the Arabs in ancient times. Its pleasant, acidic-tasting fruit was so popular that the plant's botanical and common names both point to its association with India.

54. FLAME OF THE FOREST
Butea monosperma (Lamk.) Taub.

(Family: Faboideae)

Plate 7

Hindi *Dhak, Palas*; Bengali, Marathi *Palas*; Malayalam *Pala*; Tamil *Parasa*; Gujarati *Kesudo*; Burmese *Pauk*.

Field Identification: Tree with a crooked trunk. Flowers bright orange, 3.8-5 cm, spectacularly beautiful, in great clusters, contrasting with the jet-black calyces. Pods 10-20 x 2.5-5 cm, with a single seed. Leaves large leathery, trifoliate, terminal leaflet 13-20 x 11-15 cm. **Description:** A small or medium, deciduous tree with a crooked trunk and large irregular branches. **Bark** fibrous, light brown or grey, exuding a sticky red juice which hardens into a gum. **Leaves** consisting of 3 large leathery leaflets on a common stalk 10-23 cm long, swollen at the base, terminal leaflet 13-20 x 11-18 cm. **Flowers** 3.8-5 cm long, spectacularly beautiful, bright orange-red in great clusters contrasting vividly with the jet-black velvety 1.25 cm long calyces. **Pods** 10-20 x 2.5-5 cm, thickened at the top end where the solitary seed is lodged, the rest of the pod is thin, strongly nerved, grey-silky, pale-yellowish when ripe. The ripe pods are light and scattered by the strong winds during the hot weather. **Seeds** kidney-shaped, 3.3-3.8 x 2.5 cm. 'Dhak' with pale yellow or golden-yellow or white flowers have been reported. **Distribution:** Widespread in the Subcontinent, including Myanmar and Sri Lanka. Up to 1,220 m in the Himalaya. **Phenology:** Leafless or nearly so from February to April. New leaves appear in April-May. The black flower-buds appear in January and **flowers** burst open in great clusters of vermilion in February-March, presenting a most gorgeous sight. **Pods** ripen in May-June. The flowers are fertilized by babblers, sunbirds and other birds. **Miscellaneous:** Propagation by seed which should be sown as soon as ripe. Transplanted after 2-3 years till a stem strong enough to stand unfavourable extreme heat or cold weather is developed. It is a valuable host for the lac insect. Leaves are harvested for fodder for elephants and also used as wrappers for 'bidis'. It is useful for afforestation of saline regions. The tree is ornamental only when in flower. **Etymology:** 'Flame of the Forest' is so named as the massed crowns of bright orange flowers suggest a forest in flames. *Butea* is named in honour of John Earl of Bute, an 18th century botanist; *monosperma* in Latin means one seed. The battle field of Plassey is said to have taken its name from the Palas tree.

55. INDIAN CORAL TREE
Erythrina variegata Linn.

(Family: Faboideae)

Syn. *E. indica* Lam.

Hindi, Marathi *Pangra, Pangara*; Bengali *Palita Mandar*; Malayalam, Telugu, *Badisa*; Gujarati *Panervo*.

Field Identification: Branchlets with small black conical prickles. Calyx split to the base, tip 5-toothed. Flowers large, 5-6.3 x 2.5-3.8 cm, bright scarlet 30 cm long clusters. Stamens projecting out. Pods 15-30 cm long, black, beaked and constricted between 1-12 reddish seeds. **Description:** A medium-sized tree of rapid growth with a straight trunk. Branchlets armed with small, dark coloured conical prickles up to third or fourth year. **Bark** thin, yellowish or greenish-grey, smooth, with longitudinal whitish cracks, peeling off in papery flakes. **Leaf** of 3 leaflets on 13 cm long stalk, terminal leaflet being the largest, 10-15 x 9-13 cm, lateral leaflets smaller, covered with star-shaped hairs when young. **Flowers** large, 5-6.3 x 2.5-3.8 cm, bright scarlet, growing in 30 cm long clusters at the end of branchlets, appearing before the leaves. Calyx 2.5-3 cm, split to the base, tip 5-toothed. Corolla of 5 petals, an erect oblong standard which narrows at the base into a claw, 2 small wing petals, and similar sized keel petals of a darker hue. The wing petals partially enclose the keel. Clusters of dazzling scarlet blooms with projecting stamens look gorgeous. The stamens are united into a bundle at the base, the tenth stamen is distinct and free. Pods many, 15-30 cm, black, beaked, constricted between the 1-12 seeds. Seeds 1.5-1.8 cm, oblong reddish in colour. **Distribution:** Eastern India in its sub-Himalayan tracts, Sunderbans, Andamans, Nicobars, Myanmar and the Western Ghats. Cultivated all over India. **Phenology:** Old leaves are shed in the autumn. New leaves appear after the flowers. **Flowers** February-May. **Fruit** ripens from May-July. **Miscellaneous:** A variety of birds, rosy starlings, babblers, drongo, tailor bird, bulbuls and sunbirds are visitors to the coral tree when in flower, contributing to its pollination. They visit the freshly opened flowers to drink the sweet nectar. The tree is used as a support for pepper and betel vines and the wood for dugouts. Being ornamental it is a prized tree in gardens and parks. A white variety is reported from Lanouli (Western Ghats) and was reported to be in cultivation in Victoria Gardens, Mumbai. It propagates readily from seeds or cuttings. Even large cuttings up to 2 m long root readily. **Etymology:** *Erythrina* is from the Greek *erythros* (red), referring to the colour of the flowers.

56. INDIAN KINO TREE
Pterocarpus marsupium Roxb.

(Family: Faboideae)

Hindi *Bijasal*; Marathi *Bibla*; Gujarati *Hira dakhan*; Malayalam *Venga*; Tamil *Vengai*; Telugu *Yegi*.

Fig. 28

Field Identification: Leaflets 5-7, alternate, 3.5-13 x 1.8-5.5 cm, tip notched or round. Flowers fragrant, yellow, 1-1.3 cm long in terminal clusters. Calyx 6 mm long, bell-shaped, corolla twice as long, projecting out. Pods nearly circular, beaked on one side, pale yellow, 3.5-5.5 cm across, one-seeded. **Description:** A tall deciduous tree, attaining its largest size in the moist deciduous forests of the west coast where trees up to 30 m high and 5 m girth are known. **Bark** grey, peeling off in irregular scales, blaze pink with whitish markings. Old trees exude blood-red astringent gum from the cut. **Leaves** large, 15-30 cm long with 5-7 leathery alternate leaflets, 3.5-13 x 1.8-5.5 cm, sometimes notched at the tip, or round, stalk of leaflets. **Flowers** fragrant, yellow, 1-1.3 cm long in terminal 15-25 cm long clusters. Calyx 6 mm long, bell-shaped, rusty hairy. Corolla projecting out and twice as long as the calyx, petals with long claws. Stamens 10, filaments united into a tube. **Pod** 3.5-5.5 cm long, light yellowish brown, nearly circular, beaked on one side, usually one seeded. **Seed** 1-1.3 cm, reddish brown, small and leathery. **Distribution:** In deciduous forests in the peninsula and Sri Lanka, also from Uttar Pradesh to Orissa and Madhya Pradesh. **Phenology:** Leafless for a short time during April-May. New leaves appear in May, June. **Flowers** May-August. **Pods** ripen from December-March. The strong winds disperse the

Fig. 28

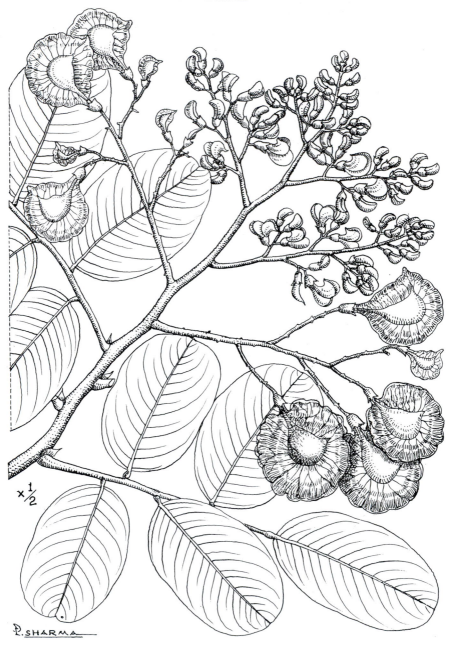

$\times \frac{1}{2}$

P. SHARMA

S.No. 56. ***Pterocarpus marsupium***. Flowers, leaflets and pods x $\frac{1}{2}$. The pod shaped like a kangaroo's pouch explains the specific name.

seeds far and wide. **Miscellaneous:** In South India the timber is ranked after teak and rosewood for furniture. The exudate from the bark is a blood-red resin called 'Kino' used in medicine and in European wines. The tree as a sapling is browsed by deer. **Etymology:** *Pterocarpus* means a winged seed, *marsupium* means the pouch of a kangaroo which the fruit resembles.

57. PADAUK, ANDAMAN RED-WOOD
Pterocarpus dalbergioides Roxb.

(Family: Faboideae)

Hindi *Padauk.*

Fig. 29

Field Identification: Large buttressed tree exuding a blood red juice when blazed. Leaflets 5-9, alternate, 5-10 cm long, pointed. Flowers yellow, 1.3 cm long, in clusters. Pods circular, winged, 5-6.3 cm across, beaked. Seeds 1-2, 1.3 cm across, reddish brown, smooth and shining. **Description:** A large, semi-evergreen or deciduous buttressed tree of the Andamans, 24-36 m high and 4.2 m in girth. Trees up to 49 m and 6 m girth are recorded. When blazed it exudes a blood-red juice and is therefore also called Andaman red-wood. Buttresses sometimes produce large burrs. **Leaves** compound, 20-25 cm long with 5-9 alternate leaflets, 5-10 cm long, egg-shaped, tapering gradually to a point. **Flowers** golden yellow, 1.3 cm long in terminal clusters. **Pod** flat, circular, winged, 5-6.3 cm wide, beaked, tapering down to the stalk, one or two seeded. **Seed** 1.3 cm long, flattened, smooth and shining, reddish brown. **Distribution:** Andamans, absent in Nicobars. Cultivated in Bengal and South India. **Phenology:** Leafless during February-March. **Flowers** May-June. **Pods** ripen December-February, reddish brown seen on the trees till May. They are dispersed far and wide by the high winds of the early monsoon. **Miscellaneous:** The unripe green pods are destroyed in large quantities by parakeets, chital browse the leaves of young saplings and cause serious damage. The buttress produces large burrs which, when sliced with power-saws, turn out beautifully figured veneers used as table tops after polishing. The timber is used in panelling and furniture and varies in colour from light brown to a gorgeous red.

FIG. 29

S.No. 57. ***Pterocarpus dalbergioides***. Flowers, leaflets and pod x 1. (From *Heinig* 90, 16).

58. RED SANDERS
Pterocarpus santalinus Linn. f.

(Family: Faboideae)

Hindi: *Lal chandan*; Telugu *Yerra chandanum*.

Fig. 30

Field Identification: Small tree with a rounded crown, compound leaves with 3 broadly elliptic leaflets, 3.8-7.6 cm, yellow flowers with stalks shorter than calyx. Pod 3.8 cm across, obliquely circular, winged with mostly one reddish brown leathery seed 1-1.5 cm long. **Description:** A small deciduous tree with a rounded crown. **Bark** blackish brown, cleft into rectangular plates by deep vertical and horizontal cracks, exuding a red gum when blazed. **Leaves** compound, 10-18 cm long, leaflets 3, rarely 4 or 5, 3.8-7.6 cm, broadly elliptic, slightly notched, hairy beneath. **Flowers** yellow, calyx 5-6 mm, with small triangular teeth. Flower stalk shorter than calyx, 5 mm long. **Pod** 3.8 cm across, obliquely circular, gradually narrowed to a short stalk. **Seeds** one or two, 1-1.5 cm long, reddish brown, smooth and leathery. **Distribution:** Endemic to Cuddapah District in Andhra Pradesh and in Chinglepet in Tamil Nadu, growing on rocky hills at 150-900 m. **Phenology:** **Leaves** are shed from January-March. New leaves appear in April. **Flowers** April-May. **Pods** ripen next February to March and remain on the tree till May and are dispersed by the south-west winds. **Miscellaneous:** Sambar and chital greedily browse the seedlings. The fragrant heartwood is claret-purple to dull black and contains a dye called santalin. The wood is used in carving and also exported to Japan for making musical instruments. The timber with wavy grain fetches a fancy price as it produces a fine resonance. Being endemic to a small area in Andhra Pradesh it is now on the endangered list of flora. In the past it has been ruthlessly felled and exploited. The common name red sandalwood is inappropriate as the tree does not belong to the Sandalwood family. As it causes confusion it is not used here.

59. ROSE WOOD
Dalbergia latifolia Roxb.

(Family: Faboideae)

Hindi, Bengali *Satsal*; Malayalam, Tamil *Itti*; Telugu *Cittegi*.

Fig. 31

Field Identification: Tree with compound leaves with 5-7 leaflets, 2.5-7.5 cm, broadly ovate, notched or round at apex. Flowers small, white, 0.63 cm. Pods 4-9 x 1.3-2 cm flat, strap-shaped. Seeds 1-3. Wood reddish brown. **Description:** A

Fig. 30

S.No. 58. *Pterocarpus santalinus*. 1) Flowers, leaflets x 1. 2) Winged pod x 1.

Fig. 31

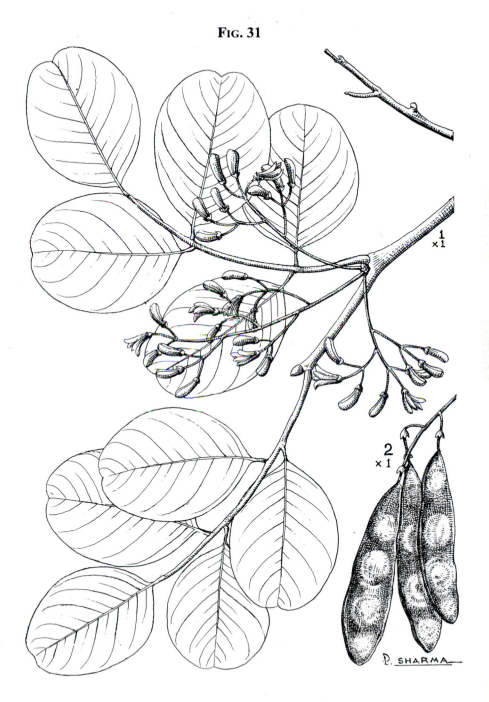

S.No. 59. ***Dalbergia latifolia***. 1) Flowers, leaflets x 1. 2) Pods x 1. (From *Robinson* 36823).

large deciduous tree attaining its best size in the Western Ghats where trees up to 40 m in height are found. **Bark** ash grey, peeling off in thin long flakes. **Leaves** alternate, 10-25 cm long, leaflets 5 (sometimes 5-7), 2.5-7.5 cm long, broadly ovate, notched or round at apex, dark olive green, downy beneath, with 5-6 pairs of lateral nerves. **Flowers** white, 0.63 cm in clusters shorter than leaves. Corolla 0.6 cm, petals clawed. Stamens 9, filaments united into a tube. **Pods** firm, 4-9 x 1.3-2 cm, flat, strap-shaped. Seeds 1-3. **Distribution:** Sub-Himalaya from Eastern U.P. to Sikkim, Central, Western and South India attaining its best dimensions on the Ghats of north Kanara. **Phenology: Leaves** are shed from February-March. New leaves in April, May. **Flowers** mostly April, flowering time varies from place to place. **Pods** ripen December-February. **Miscellaneous:** The sapling is browsed by deer and gaur. The gold-brown to rose purple or deep purple heartwood, streaked with black, ranks among the finest woods for furniture and is a prized timber for making musical instruments e.g. pianos, clarinets and guitars.

60. SISSOO
Dalbergia sissoo Roxb.

(Family: Faboideae)

Hindi *Shisham*; Punjabi *Tali.*

Fig. 32

Field Identification: Tree with dark grey rough bark. Leaves of 3 alternate broad-ovate leaflets, 2.5-6.3 cm across, on 3-5 mm long stalks. Flowers small, 5-8 mm, yellowish-white, inconspicuous. Pod strap-shaped, 3.7-7.5 cm x 8-11 mm, 1-4 seeded. **Description:** A fairly large deciduous tree with dark grey, rough and furrowed **bark. Leaves** composed of 3 leaflets arranged alternately on a zig-zag axis. Leaflets broad-ovate, 2.5-6.3 cm, pointed, on 3-5 mm long stalks. **Flowers** small, 5-8 mm, yellowish-white, numerous, inconspicuous but scented. Stamens 9, united in a tube slit along the upper side. **Fruit** a thin strap-shaped pod, 3.7-7.5 cm x 8-11 mm, with 1-4, kidney-shaped light brown **seeds** 6-8 x 4-5 mm broad. **Distribution:** Sub-Himalayan tracts of the Subcontinent, growing along river beds. **Phenology:** Leafless in January-February. New foliage appears in March. **Flowers** March. **Fruit** ripens in November but remain hanging on the trees for several months. **Miscellaneous:** Parakeets destroy quantities of pods when still green. It is the most extensively planted timber tree after teak. Fast growing, adaptable and able to stand temperatures, from below freezing to 50°C. It is therefore used to shade bushes and is also a high class furniture wood. **Etymology:** *Dalbergia* is named in honour of Nicholas Dalberg, a 19th century Swedish botanist, *sissoo* is one of the vernacular names for the tree.

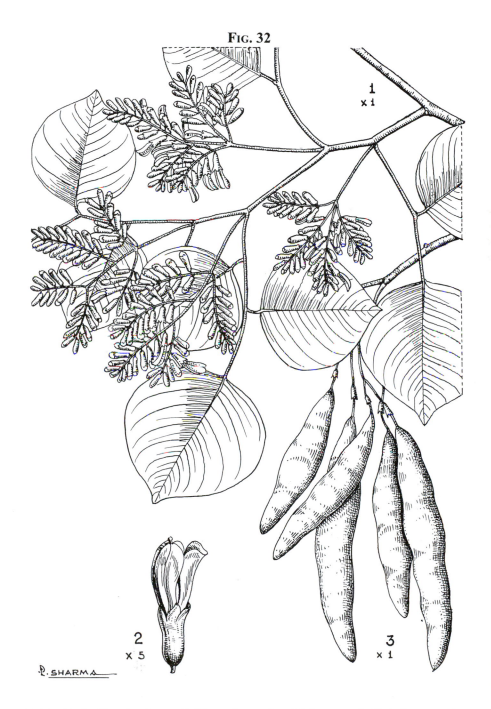

Fɪɢ. 32

1
x 1

2
x 5

3
x 1

P. SHARMA

S. No.60. ***Dalbergia sissoo***. 1) Flowering twig with leaves x 1. 2) A flower x 5.
3) Pods x 1.

61. PONGAM OIL TREE
Pongamia pinnata (L) Pierre
(Family: Faboideae)

Hindi, Bengali, Marathi, Gujarati *Karanj, Karanja*; Tamil *Ponga*; Malayalam, Telugu *Pungu*; Burmese *Thinwin*.

Fig. 33

Field Identification: Mostly a low tree with an umbrella-shaped crown. Leaves of 5,7 or 9 leaflets. Flowers white, tinged with pink or violet, 1.3 cm long, in conical clusters. Pods woody, not opening when ripe, 3.8-5 x 1.8-2.5 cm, pointed at the tip, with mostly 1 reddish seed. **Description:** A medium sized, nearly evergreen tree with short trunk and umbrella-shaped shady crown. **Bark** smooth grey. **Leaves** of 5, 7 or 9 opposite leaflets 5.8 cm long, egg-shaped to oblong, abruptly pointed. **Flowers** 1.36 cm long on 0.6 cm long stalks, white, tinged with pink or violet in cone-shaped clusters. Calyx bell-shaped, 0.37 cm long. Petals 1.3 cm long, fused at the tip. **Pod** 3.8-5 x 1.8-2.5 cm, woody, pointed at the apex, yellowish grey when ripe, ultimately dark grey, not opening when ripe. **Seeds** mostly 1, sometimes 2, reddish brown. **Distribution:** Sandy beds of streams and on the sea coast of Western Ghats, Andamans, Nicobars, Myanmar and Sri Lanka. Now wild in most parts of the Subcontinent but has undoubtedly spread from planted trees. Capable of growing far inland. **Phenology: Flowers** April-June. **Pods** ripen March-May of the following year. **Miscellaneous:** Mainly used as a shade tree in avenues in New Delhi, Dehra Dun and elsewhere. The seeds give a reddish brown oil used as an illuminant and as a lubricant for engines. **Etymology:** *Pongamia* is from the Tamil name pongam or ponga; *pinnata* in Latin means leaflets arranged on either side of the stalk.

62. SANDAN
Ougeinia oojeinensis (Roxb.) Hochreut
(Family: Faboideae)

Syn. *O. dalbergioides* Benth.

Hindi *Sandan*; Marathi *Tiwas*.

Field Identification: Bark like crocodile skin, exuding a red gum when cut which turns purple on the knife blade. Leaves trifoliate. Flowers lilac in dense round clusters 2.5-5 cm across, borne on the old wood, appearing like corals. Pod linear-oblong, 5-10 x 0.80 cm. Seeds 2-5, 1 cm wide, smooth, brown. **Description:** A middle sized, deciduous tree often with a crooked trunk. **Bark** like crocodile skin,

FIG. 33

S.No. 61. ***Pongamia pinnata***. 1) Leaves, inflorescence x 3. 2) The largest of the petal x 3. 3) Pod x 1. (From *Parker* 38644).

fissured horizontally and vertically, which when blazed reveals blood-red streaks and exudes a red gum similar to 'Kino', turning purple on the knife blade. Leaves of 3 leaflets, on 5-15 cm long stalk, terminal leaflet largest, 7.5-15 x 5.8 cm, broadly egg-shaped, elliptic or rhomboid, laterals smaller. The pink or lilac **flowers** are produced in dense round clusters 2.5-5 cm across from the old wood and look like corals, flower stalks slender 1.3-2 cm, calyx 2.5-3.8 mm. Corolla 0.8-1 cm, standard, nearly circular, shortly clawed, wings oblong, keel obtuse. **Pods** linear-oblong, flat, 5-10 x 0.8 cm. **Seeds** 2-5 cm long, smooth and brown. **Distribution:** Common in the Sal forests of Uttar Pradesh. From the Ravi to Bhutan through Sikkim, Central and South India. **Phenology:** Old **leaves** turn yellow before falling and are shed in January-February. New leaves appear in March-April and are copper-red. **Flowers** when leafless February-April. **Pods** ripen May-June. **Miscellaneous:** It is a spectacular sight when in bloom in March and therefore worthy of introduction in parks and gardens. Leaves are good fodder for cattle. The bark is used to intoxicate fish. **Gardening:** Raised from seed which is sown soon after collection after drying in June. Germination takes place early during the monsoon.

63. JEWELS ON A STRING, MOULMEIN ROSEWOOD
Milletia ovalifolia Kurz
(Family: Faboideae)

Burmese *Nantha*.

Field identification: Tree with drooping branches. Leaflets 7, 6 in 3 opposite pairs and 1 terminal, egg-shaped, varying in size from 4-8 x 4.5 cm. Flowers in drooping sprays with steel-blue or magenta red petals, standard petal 0.6 cm long. Pod elliptic, 5-7.5 cm, narrow at base and pointed at apex, almost woody, flat. Seeds 2-3. **Description:** A small to medium sized tree 12-15 m tall with a rounded crown and drooping branches. The elegant drooping **foliage** consists of 7 egg-shaped leaflets arranged in 3 opposite pairs on either side of a slender midrib, with a terminal leaflet at the tip. The compound leaf is about 20-24 cm long. Leaflets vary in size from 4-8 x 4.5 cm on 0.5 cm long stalks. **Flowers:** petals steel-blue or magenta rose in colour, sepals garnet-like in colour. The two together produce a 2 colour effect. The standard petal is 0.6 cm long. The small flowers hang in 5-7.5 cm long drooping sprays. **Pods** elliptic, 5-7.5 cm, narrow at the base and pointed at the apex, almost woody, flat, covered with minute warts. **Seeds** 2-3. **Distribution:** Myanmar, dry forests of Prome and entering the Savanna forest. **Phenology:** Leafless from December to February. New **leaves** soon after flowering in March-April. **Flowering** and **fruiting** March-April.

Miscellaneous: Grown from seed. It is a spectacular sight when in bloom and is now commonly seen in parks in Delhi, Lucknow and elsewhere. **Etymology:** *Milletia* is named after French botanist J.A. Millet; *ovalifolia* in Latin means egg-shaped leaves. It is one of the most beautiful trees when in bloom, with delicate sprays of steel-blue flowers before the leaves appear.

64. WILD PEAR
Pyrus pashia Buch.-Ham. ex D. Don
(Family: Rosaceae)

NW. Himalaya *Mehal, Mol*; Burmese *Sakhaw.*

Field Identification: A small tree, with almost black bark and spiny stems. Leaves egg-shaped, pointed, 5-8 x 2-4 cm, toothed. Flowers white, 2-2.5 cm across. Stamens many. Fruits round, 1.3-2.5 cm, dark brown with white specks. **Description:** A small or medium sized deciduous tree. **Bark** almost black on old stems. Barren stems sometimes spiny. **Leaves** egg-shaped to broadly lance-shaped, pointed, 5-8 x 2-4 cm, toothed, shining when mature. **Flowers** 2-2.5 cm across, white, at ends of branchlets in clusters, calyx urn-shaped, petals rounded, stamens 25-30. **Fruits** globose, 1.3-2.5 cm, dark brown when ripe, with raised white specks on the surface. **Distribution:** All along the Himalaya from Pakistan to Bhutan, Khasi Hills, Manipur and Myanmar at 650-2500 m. **Phenology: Leaves** appear in March-April after flowering. **Flowers** February-April, **Fruits** September-December. **Miscellaneous:** Fruit eaten when half rotten. Leaves lopped for fodder. Wood used for making walking sticks.

65. TRUE MANGROVE
Rhizophora mucronata Lamk.
(Family: Rhizophoraceae)

Marathi *Kamo*; Sind *Kandal*; Telugu *Uppu, Poma*; Bengali *Bhara*; Andamans *Burada*; Burmese *Pyu.*

Fig. 34-1

Field Identification: Tree with stilt-roots. Leaves broad, elliptic with black dots, 7.5-20 x 4.5-10 cm, terminated by a small sharp tip at the apex. Flowers yellowish white, nearly 2.5 cm across. Fruit dark brown with a radicle 30-60 cm long. **Description:** A small or moderate-sized non-buttressed evergreen tree, supported on stilt-roots. **Bark** smooth with vertical clefts. **Leaves** elliptic, wedge shaped at the lower end and terminated by a small sharp tip at the apex called *mucro* in Latin

FIG. 34

1

$\times \frac{1}{2}$

2

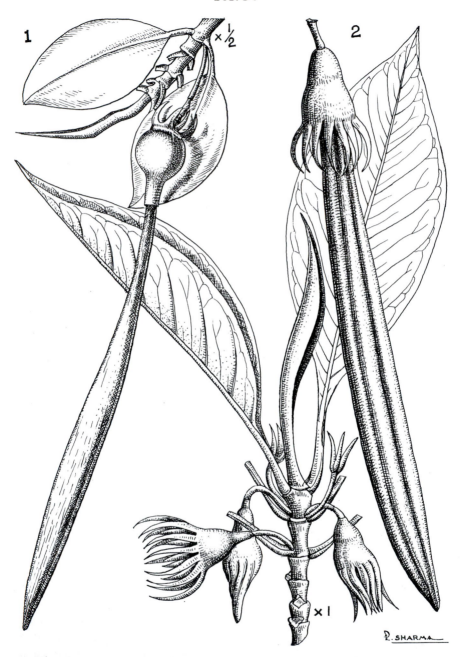

S.No. 65. **Rhizophora mucronata** 1) Leaves, fruit x ½. S.No. 66. **Bruguiera gymnorhiza** 2) Leaves, flowers, fruit x 1. (From *K.C. Sahni* 22902. Gt. Nicobar Is.).

and hence the specific name. Leaves 7.5-20 x 4.5-10 cm, thick, bright green above, pale beneath dotted with minute dark spots; petiole, 1.8-3.8 cm. **Flowers** in threes, yellowish-white 2.5 cm across. Petals oblong, thick, fleshy. Stamens 8. **Fruit** 3.8-5 cm long, dark brown; radicle (embryo roots) 45-60 cm long, before falling from the tree. Seeds germinate on the parent tree and drop into the mud as young plants. **Distribution:** Indus to Myanmar, along both East and West coast of India and backwaters. Sunderbans, Andamans, Nicobars and Sri Lanka. In the Andamans and Nicobars, it forms impenetrable belts of vegetation at the water's edge many kilometres long and 1.5 km in depth. **Phenology: Flowers** during the rainy season. **Fruits** Aug-Sep. Flowers and fruit are also seen all the year round. **Miscellaneous:** The bark is used in tanning. Wood greatly valued as fuel. The fruit is sweet and edible.

66. MANGROVE
Bruguiera gymnorhiza (L.) Lamk
(Family: Rhizophoraceae)

Vernacular *Kankara.*

Fig. 34-2

Field Identification: Buttressed tree. Leaves thick, leathery, opposite, 6.3-10 x 2.5-3.75 cm, tapering at both ends. Flowers solitary, orange and red, calyx lobes 10-14, linear-erect, 2.5-3.7 cm, petals 8-14, deeply cleft. Stamens 16-28 in pairs. **Fruit** top-shaped, 2 cm long, crowned by scarlet calyx segments, radicle 20 cm long x 1.25 cm wide, faintly ribbed, brown and pointed. **Description:** An evergreen buttressed tree over 20 m tall, one of the largest of the mangroves. **Bark** dark, fissured. **Leaves** thick, leathery, elliptic-lance-shaped, tapering at both ends 6.3-10 x 2.5-3.75 cm, leaf stalk 1.2-1.8 cm long. **Flowers** large, solitary, orange-red. Calyx lobes 10-14, linear, erect, 2.5-3.7 cm long, orange-yellow, petals 8-14, deeply cleft. Stamens 16-28, in pairs opposite petals, springing elastically from them when mature. **Fruit** top-shaped, 2 cm long, crowned by the scarlet calyx segments which fall off; radicle brown 20 x 1.25 cm, faintly ribbed and pointed. **Distribution:** From the Indus delta in Pakistan along the Indian coast, Bangladesh, Andamans and Nicobars, Sri Lanka and Myanmar in littoral swamps, creeks and tidal rivers. **Phenology: Flowers and Fruits** during the rainy season. **Miscellaneous:** An excellent fuel wood, also used for house posts and furniture.

67. ARJUN
Terminalia arjuna Wight et Arn.
(Family: Combretaceae)

Hindi *Arjun*; Gujarati *Dhaula, Sad*; Kannada *Holematti*; Telugu *Thella maddi.*

Fig. 35

Field Identification: Large buttressed tree with drooping branches. Leaves leathery, hard, elliptic, 7.5 cm long. Flowers small, pale yellowish white, without petals, on 5-7 cm long spikes near tips of branches. Fruit 5-winged, 2.5-3.7 cm long, woody. **Description:** A large evergreen tree, often buttressed at the base. **Bark** smooth, grey, flaking off in thin layers. Branches drooping. **Leaves** opposite, or sub-opposite, oblong or elliptic, hard and leathery, 7.5-15 cm long, with a pair of glands near the top of the leaf-stalk. **Flowers** pale yellowish white, small but crowded on 5-7.5 cm long axillary spikes near the tips of branches, petals absent, stamens 10. The **fruit** is a winged nut 2.5-3.7 cm long, the wings usually 5-7 in number, about 1.2 cm wide and closely veined. **Distribution:** Central and South India to Sri Lanka along streams. It has been introduced into North India. **Phenology:** Evergreen or nearly so. New foliage appears early in the hot season. **Flowers** April to July. **Fruit** ripens February-May. **Miscellaneous:** A popular shade tree, planted in avenues and parks. Bark used in tanning leather. Leaves are fed to silkworms. The wood is used in buildings and in boats. **Etymology:** *Terminalia* is from the Latin *terminalis,* meaning terminal. So named because the leaves are in tufts at the tips of branches in many species (though not in this tree); *arjuna* is the latinized Hindi name for the tree.

68. HOLLOCK
Terminalia myriocarpa Heurck & Muell.-Arg.
(Family: Combretaceae)
Plate 8

Nepali *Panisaj*; Assamese *Hollock*; Khasi. *Dieng-tal.*

Field Identification: A gigantic buttressed tree of the E. Himalayan foothills with drooping branches. Covered with masses of tiny creamish flowers in autumn, which turn coppery red on fruiting in November. Fruit 0.4-0.5 cm, 3 cornered in drooping panicles. **Description:** A large evergreen tree up to 35 m tall, buttressed at the base, with long drooping branches. **Bark** greyish brown, peeling off in long flakes. Blaze salmon turning brown. **Leaves** opposite, elliptic-oblong, 10-25 x 4-7 cm, with a toothed margin, lateral nerves numerous and parallel leaf stalk 0.7

FIG. 35

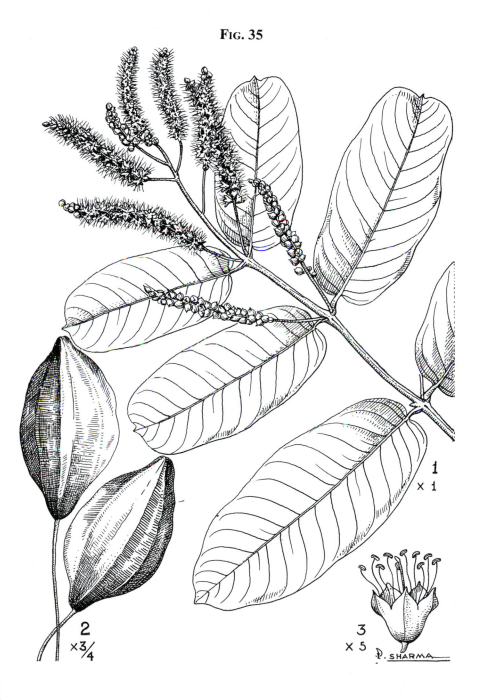

S.No. 67. **Terminalia arjuna.** 1) Leaves and inflorescence x 1. 2) Fruit x $\frac{3}{4}$.
3) Flowers x 5.

cm long, with one or two prominent glands at the top. **Flowers** small, yellow, 0.3-0.4 cm long on slender hanging panicle up to 30 cm long. Stamens 10. **Fruit** 0.4-0.5 cm long, 3-cornered, the two lateral angles expanded into wings. The tree is very handsome when covered with masses of tiny creamish flowers in autumn. These turn coppery red in November on fruiting, presenting a most spectacular sight. **Distribution:** Eastern Himalaya from foothills of Nepal, Darjeeling, Sikkim, Arunachal Pradesh up to 1400 m. Also in Assam, Khasi Hills and Manipur, Bhutan, Myanmar. **Phenology: Flowers** September-October. **Fruits** November. **Miscellaneous:** A very handsome tree worthy of introduction in parks in subtropical climates. It does extremely well in Dehra Dun where some excellent trees are flourishing in the Forest Research Institute. A valuable plywood timber, much in demand for making tea chests. The wood takes a beautiful polish. Raised from seed, sown in fine porous sandy soil in boxes or in well raised beds protected from sun and heavy rain. Watering should be frequent, but light. Germination starts in 2 or 3 weeks. Transplanted within the first season when 7-10 cm high.

69. LAUREL
Terminalia crenulata Heyne ex Roth.
(Family: Combretaceae)

Vernacular *Karumaruthu, Matti, Tehmbava*, West Coast; Bengal *Paka Saj.*

Field Identification: A large tree. Bark dark-grey, furrowed, peeling off in rectangular flakes. Leaves subopposite or alternate, 18 x 7 cm, margin round, toothed, with 2 glands from lower midrib. Flowers pale yellow on a branched spike. Fruits coppery-red, 5 winged, 3 x 3.5 cm. **Description:** A large deciduous tree with dark-grey furrowed **bark** peeling off in rectangular flakes. **Leaves** subopposite or alternate, round, toothed along the margins (which gives the tree its specific name), elliptic-obovate, 18 x 7 cm, with 2 stalked glands from lower midrib. **Flowers** pale yellow on a branched spike, calyx not hairy, stamens 10. **Fruit** 5-winged, coppery-red, rounded at the top, marked with straight striations at right angles to the axis, 3-3.5 cm long. **Distribution:** South-west and Central India, North Bengal, Assam (Goalpara); common in the Western Ghats, Myanmar. **Phenology: Flowers** July-August. **Fruits** February. **Miscellaneous:** The best figured laurel is found on the west coast from where it is exported. It is used for decorative panelling, high class furniture and cabinet work. The wood is dark brown streaked with black and is also good for building.

70. MYROBALAN
Terminalia chebula (Gaertn.) Retz.

(Family: Combretaceae)

Gujarati *Harda*; Hindi *Harra*; Bengali *Haritaka*; Marathi *Hirda*;
Tamil *Kadukkai*; Telugu *Karakkai*; Burmese *Panga*.

Fig. 36

Field Identification: Leaves alternate or opposite, elliptic-egg-shaped, 9-20 x 3.5-8 cm, leathery, with a pair of glands on top of the 1.2-2.5 cm long leaf stalk. Flowers small, 0.3 cm across, creamish with an offensive smell, borne in long terminal panicles. Fruits hanging down, ellipsoid-egg-shaped, 2.5-5 cm long, yellow to orange brown, 5-ribbed on drying, stone bony and angled. **Description:** A medium-sized deciduous tree with a short crooked trunk. **Bark** dark brown with vertical cracks. **Leaves** alternate or opposite, elliptic or egg-shaped with a short pointed tip, 9-20 x 3.5-8 cm long, leathery, with a pair of large glands on the top of the 1.2-2.5 cm long leaf stalk. **Flowers** small, 0.3 cm across, bisexual, whitish or yellowish with an offensive smell, usually in terminal panicles in the axils of the uppermost leaves; calyx cup-shaped, cleft into 5 triangular segments, petals absent, stamens 10, inserted on the calyx tube. **Fruit** hanging down, ellipsoid or egg-shaped, 2.5-5 cm long, yellow to orange brown, sometimes tinged with red or black, on drying 5-ribbed, stone hard, bony and angled. **Distribution:** Sub-Himalayan tracts. Deciduous forests of the Western and Eastern Ghats. It is plentiful in the hill resort of Mahabaleshwar and in the Kangra valley. In the latter it grows mixed with 'chir' pine. **Phenology: Leafless** by February or March. New **leaves** appear from March to May. **Flowers** March-May. **Fruit** ripens from November to March. **Miscellaneous:** Fruit yields the most important tanning material of the Subcontinent for dyeing leather, wool and cotton. Wood is durable and used in house building.

71. WHITE CHUGLAM, SILVER-GREY WOOD
Terminalia bialata Steud.

(Family: Combretaceae)

Andaman & Nicobar Is. *Safed chuglam*; Burmese *Lein*.

Field Identification: A very tall tree easily recognized by its fallen fruits on the forest floor which are butterfly-shaped, with 2 broad wings, 5 x 10 cm, light brown in colour. **Description:** A very large and tall tree up to 47 m with large, thin, curved and often branched buttresses. **Bark** light brown, smooth and finely fissured.

Fig. 36

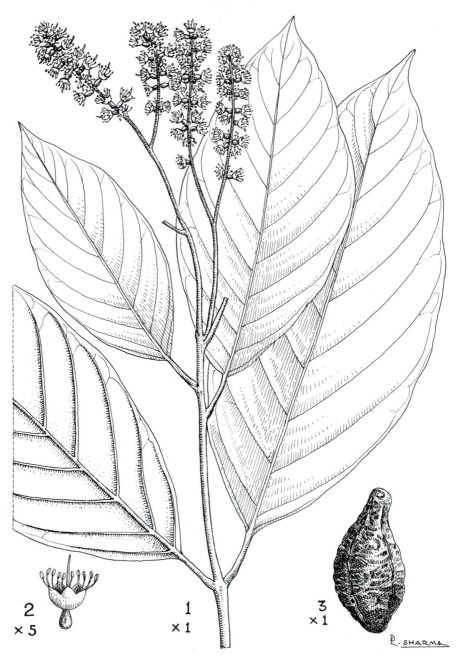

S.No. 70. **Terminalia chebula** 1) Leaves and inflorescence x 1. 2) A flower x 5.
3) Fruit x 1.

Leaves alternate, crowded at ends of branchlets 15-22.5 x 7.5-10 cm, obovate or oblanceolate, shortly pointed, tapering at base; leaf stalk 5-7.5 cm without glands unlike other *Terminalia* species which have glands. **Flowers** small yellowish white in simple axillary spikes, as long as leaves, the upper flowers male, the lower bisexual, ovary and calyx densely hairy. **Fruit** butterfly-shaped, 5 x 10 cm, with 2 broad stiff, veined wings light brown in colour. **Distribution:** Andamans, Nicobars and Myanmar. One of the largest and most magnificient trees of the Andamans and Nicobars, it does not attain the girth of Dipterocarps but surpasses them in height. **Phenology: Leafless** in hot season. New **leaves** and **flowers** May-June. **Fruit** fall in February-April. **Miscellaneous:** The timber is grey, beautifully mottled, strong and elastic and takes a fine polish. Used in furniture and planking in house construction.

72. JAMUN, BLACK PLUM
Syzygium cumini (Linn.) Skeels
(Family: Myrtaceae)

Hindi *Jaman*; Marathi *Jambul*; Bengali *Jam*; Nep. *Kala Jam*; Gujarati *Jambu*, Malayalam *Nhaval*, Telugu *Neereedu*; Tamil *Neredam*, Burmese *Thabye*.

Fig. 37

Field Identification: Leaves opposite, gland-dotted, with an intra-marginal vein. Flowers small, greenish white and fragrant. Petals united into a cup which falls off. Fruit globose-oblong, 1.25-3.7 cm, purplish-black when ripe, with a juicy edible pulp. **Description:** A medium sized evergreen tree with a dense shady much-branched crown. **Bark** thick, smooth, light grey. **Leaves** opposite, gland-dotted (dots visible when held against the sun), aromatic, smooth, leathery, shining, 7-15 cm long, leaf stalks 1.2-2.5 cm. The leaves have a characteristic vein pattern helpful in identification. Lateral veins are very fine, parallel and run straight from the midrib towards the margin and unite with the intra-marginal vein which runs round the leaf close to the leaf margin. **Flowers** small, greenish white, fragrant, stalkless and appear from the twigs usually below the leaves, crowded in small heads; calyx tube top-shaped, 0.2-0.5 cm, petals united into a cup or calyptra and fall off in one piece, stamens many. **Fruit** globose or oblong, 1.25-3.7 cm long, purplish-black when ripe, with a juicy edible pulp. **Seed** one in each fruit, 1-2 cm long, shaped like the fruit. **Distribution:** Common throughout the Subcontinent, except the semi-arid parts of Sind, Rajasthan and Punjab. **Phenology:** The **leaves** commence falling from January to March. New leaves which are coppery red in colour, appear in February-March. **Flowers** March-May. **Fruit** ripens from June to August. Fruits are devoured by flying foxes and birds and seeds thus dispersed.

FIG. 37

S.No. 72. *Syzygium cumini*. 1) Leaves, flowers x 1. 2) Fruits x 1. 3) Flower in section x 5.

73. QUEEN'S FLOWER TREE, QUEEN'S PRIDE OF INDIA, JARUL
Lagerstroemia reginae Roxb.

Syn. *L. flos-reginae* Retz.

(Family: Lythraceae)

Hindi, Bengali *Jarul*; Assamese *Aghar*; Marathi *Taman*; Tamil *Pumarathu*; Malayalam *Manimarathu*; Burmese *Pyinma*.

Field Identification: Tree with a rounded crown with great conical clusters up to 30 cm long laden with mauve flowers 5-7.5 cm across. Flowers with petals crinkled like crepe-paper and hence the popular name crepe myrtle. Leaves opposite, large 12.5-20 x 3.7-7.5 cm, with prominent lateral nerves. **Description:** A medium sized deciduous tree with a rounded crown growing to a large size when along banks of streams. **Bark** smooth, greyish, peeling off in irregular flakes. **Leaves** opposite or subopposite, leathery, oblong-lance-shaped with a pointed tip, 12.5-20 x 3.7-7.5 cm on 0.6-1.25 cm long stalks. Lateral nerves prominent and curving upwards. The tree is covered with 30 cm long conical clusters of mauve **flowers**, 5-7.5 cm across with 6-7 petals crinkled like crepe-paper and heavily 12 ribbed calyces. The petals are clawed at the base. Sepals 6-7. Stamens are purplish red with yellow anthers. **Capsule** broadly egg-shaped calyx. **Seeds** 1.5-1.8 cm long, light brown with a stiff brittle wing, thin and light. **Distribution:** Western Ghats along banks of streams, Bengal, Bangladesh, Assam, Myanmar and Sri Lanka. **Phenology: Leaves** are shed about February-March, and turn reddish before falling. New leaves appear in April-May. **Flowers** April-June. **Fruit** ripens November-January. They dehisce and shed seeds about February-March. Fruits are at first green but later turn brown and finally black and hang on the tree for a long time. **Miscellaneous:** Grows from seed. Transplanting is done when the seedlings are a year old. Flowers 3-5 years after planting. A highly ornamental tree; the timber is ranked next in value after teak in Myanmar. Used for boat building and furniture. **Etymology:** *Lagerstroemia* is named after Magnus Lagerstroem (1691-1759), a Swedish merchant who received specimens from the east and sent them to his botanist friend Linnaeus (1696-1759) for naming; *reginae* means imperial or Queen's Flower. It is the state flower of Maharashtra. Jarul is one of the most striking trees of the damp jungles of Eastern India and the Western Ghats. It is not often that a tree with handsome flowers has valuable timber. This species has a strong timber which is valuable.

74. JUNGLI DUNGY
Tetrameles nudiflora R. Br.

(Family: Tetramelaceae)

Hindi, Andamans *Jungli Dungy*; Tamil *Chini*; Kannada *Yermal*; Malayalam *Vella Pasa*; Marathi *Jungli-bhendi*; Bengali *Sandugaza*; Assamese *Bhelu*; Khasi *Dieng sharat*; Burmese *Thitpok*.

Field Identification: Enormous tree with huge wall-like buttresses. Bark grey, smooth and shining. Leaves circular to subcircular, 8-14 x 6.5-15 cm obscurely 5-7 angled, toothed, 5-nerved at the heart-shaped base. Flowers small, petals 0, male and female on different trees. Male flowers in erect terminal panicles, stamens 4, anthers large, white. Female flowers in hanging racemes. Fruit urn-shaped, 0.5 cm long, 8-ribbed, crowned by persistent calyx and styles, seeds minute. **Description:** A very tall deciduous tree up to 50 m, towering over other trees, with enormous wall-like plank buttresses, a long clear bole and spreading crown. **Bark** lead-grey, smooth and shining, is unmistakeable. Outer layer papery. Cut creamy white with colourless sap. **Leaves** 8-14 x 6.5-15 cm, circular or subcircular, pointed at the apex, obscurely 5-7 angled, irregularly toothed, 5 nerved at the heart-shaped base, smooth above and hairy beneath, on stalks 5-10 cm or longer. **Flowers** small, yellowish, petals 0, male and female on different trees. **Male** flowers 0.2 cm across, in erect, terminal 25 cm long panicles. Calyx lobes, linear-oblong, stamens 4, filaments twice as long or longer than calyx, anthers large, white. **Female** flowers in hanging racemes, calyx urn-shaped, 4 cleft. **Fruit** 0.5 cm long, urn-shaped, 8-ribbed, crowned by the persistent calyx segments and persistent styles. **Seeds** minute, ellipsoid and flattened. **Distribution:** Eastern sub-Himalayan tracts and outer hills from Darjeeling, Sikkim foothills to Arunachal Pradesh, Meghalaya, Myanmar, Andamans and Nicobars, Western Ghats (rainforests), Nilgiris to Sri Lanka. **Phenology:** Leaves are shed in January. **Flowers** February-April while leafless. **Fruits** April-May. **Miscellaneous:** The flowers are much sought after by bees which build their hives on the tree. A favoured nesting tree of hornbills. Woodpeckers hollow out the soft wood and nest in the holes arranged in a vertical line. It is an extremely fast growing tree used in the plywood and match industries, and to make dugout canoes in Andamans and South India.

75. STRAWBERRY TREE
Benthamidia capitata (Wall.) Hara
(Family: Cornaceae)

Syn. *Cornus capitata* Wall.

Himachal *Thama*; Garhwal *Bhamora*; Lepcha *Tumbuk*; Khasi
Dieng-sohjaphon.

Field Characters: Small tree, flowers in dense heads 1-1.5 cm, surrounded by 4
large creamish petal-like bracts 2.5 x 2 cm. Fruit red, uniting into a fleshy
strawberry-like head 2.5-5 cm across. **Description:** A small deciduous tree, with
leathery, entire, oblong or elliptic pointed **leaves** 4-7.5 cm, minutely hairy beneath,
leaf stalk 1.2 cm long. **Flowers** greenish-yellow, very small, closely packed into
hemispherical heads, 1-1.5 cm across, which are surrounded by 4 or rarely 5
conspicuous white or cream, egg-shaped, petal-like bracts 2.5 x 2 cm. **Fruit** yellow
to red, uniting into a fleshy strawberry like head 2.5-5 cm across, each drupe with
a hard 1-seeded stone, bony, compressed and angular. **Distribution:** Himachal
Pradesh to Myanmar all along the outer Himalaya, Khasi and Naga Hills at 1200-
3400 m. **Phenology: Flowers** April-October. **Fruits** November-January.
Miscellaneous: Ornamental Himalayan tree, fruits edible, used for preserves.
Etymology: *Benthamidia* is named after George Bentham, the great British botanist
of the 19th century, *capitata* refers to flowers in heads or the fruit which is a
strawberry-like head.

76. KADAM TREE
Anthocephalus sinensis (Lamk.) A. Rich. ex Walp
(Family: Clusiaceae)

Syn. *A. cadamba* Miq.

Hindi *Kadam*; Bengali *Kadamba*; Assamese *Roghu*; Telugu
Kadambe; Burmese *Mau*.

Field Identification: Branches horizontal. Leaves large, shining, opposite, elliptic
oblong, 15-30 x 11 cm, with prominent lateral nerves and linear interpetiolar stipules
which are shed early. Golden balls of yellow flowers in rounded heads a little
smaller than a golf ball. **Description:** A large deciduous tree. **Bark** grey, smooth,
becoming darker and developing longitudinal fissures when old. Its horizontal
branches, with large, shining, elliptic oblong, pointed and opposite **leaves,** 15-30
cm x 11 cm, with prominent parallel nerves, are characteristic. Leaf stalks about
2.5 cm long with linear stipules in between, called interpetiolar stipules, which
fall off early. This graceful tree is admired for its golden balls of **flowers**. The

small, scented flowers are combined in rounded heads a little smaller than a golf ball. The **fruit is** a large round orange mass of closely packed compressed angular capsules that are few **seeded** and crowned by persistent calyx lobes. **Distribution:** Sub-Himalayan tracts from Nepal to Arunachal Pradesh, Bangladesh, Bhutan, Assam, Andamans and Nicobars, Western Ghats. **Phenology:** Leafless or nearly so in the hot season. **Flowers** May-July. **Fruit** ripens and falls in January-February (Bengal Duars), and the Western Ghats. **Flowers** December-March. **Fruits** during the rainy season. **Miscellaneous:** The acidic but pleasantly flavoured fruit is relished by monkeys, bats and birds which also help in disseminating its minute seeds. The tree is associated with Lord Krishna and is sacred to the Hindus. The flowers are offered in temples. Women decorate their coiffures with Kadam flowers which have a delicate scent. It is a remarkably fast growing tree, growing up to 3 m a year, and is valued for matchwood and plywood. A spirit is distilled from the flowers. Propagated by direct sowing from March-May in raised nursery beds sheltered from sun and rain. The seedlings are pricked out when about 5 cm high. Being a large tree, it is ideal for parks and large gardens.

77. HALDINA
Adina cordifolia (Roxb.) Hook.f. ex Brandis

(Family: Rubiaceae)

Hindi, Gujarati *Haldu*; Marathi *Heddi*; Tamil, Malayalam *Manja-Kadamba*; Telugu *Pasupa-Kadamba*; Bengali *Petpuria*.

Fig. 38

Field Identification: Leaves opposite, circular, heart-shaped at the base and shortly pointed at the tip, 10-20 cm on 7.5-10 cm long stalks. Stipules whitish, covering the leaf buds and falling off early. Flower heads yellow, round, 2.5 cm across on 5-10 cm long stalks, 1-3 arising from the leaf axil. Corolla funnel-shaped with 5 bent lobes. Fruit head a collection of many capsules enclosing small winged seeds. **Description:** A large deciduous tree with horizontal branches and buttressed base. **Bark** grey, soft with great horizontal wrinkles peeling off in thick scales. Young parts hairy. **Leaves** circular, shortly pointed at tip and heart-shaped at the base, 10-20 cm, on 7.5-10 cm long stalks. Stipules whitish, up to 2 cm long, covering the uppermost pair of leaf buds, falling off early. **Flower** heads yellow, round, 2.5 cm across, on 5-10 cm long stalks, 1-3 from one leaf axil, flowers densely hairy. Calyx 5-angled, corolla funnel-shaped, 0.6 cm, with 5 bent lobes, yellow. Stamens 5, on mouth of the corolla. Stamen club-shaped. **Fruit** head a collection of numerous small capsules (4 mm long), with

FIG. 38

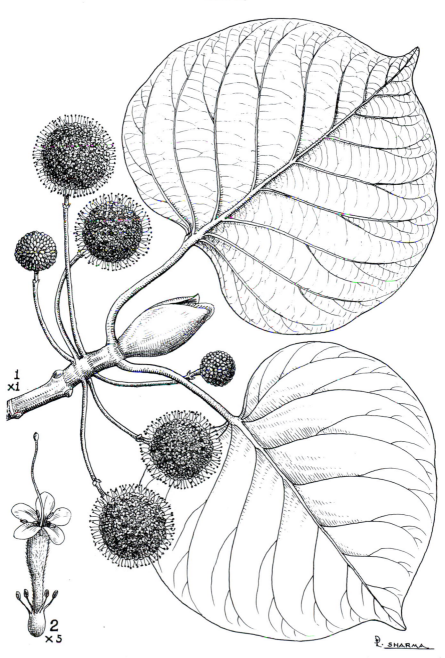

S.No. 77. ***Adina cordifolia***. Stipules, leaves and globose flower heads x 1. 2) Flower x 5. Note club-shaped stigma, inferior ovary (From *Shankamani* s.n. and *Gamble* 1500).

numerous small winged **seeds**. **Distribution:** Deciduous forests of tropical India; sub-Himalayan tracts from the Jamuna eastwards up to 800 m Western Ghats, Bangladesh, Myanmar and Sri Lanka. **Phenology:** Leafless in May-June. New **leaves** apear in June. **Flowers** June-August. **Fruit** ripens and sheds seeds from April-June of the following year. **Miscellaneous:** A very good timber for flooring and panelling. Excellent for bobbins and suitable for battery separators.

78a. TREE RHODODENDRON
Rhododendron arboreum Smith

(Family: Ericaceae)

Hazara, Pakistan, *Chahan*; Himachal *Buras*; Garhwal *Burars*; Nepal *Gurans*; Burmese *Zalutri*.

Field Identification: Evergreen tree. Leaves crowded at ends of branches, 10-15 cm long. Oblong, pointed, leathery, rusty with silvery scales beneath. Flowers blood-red in 10 cm wide clusters with *c.* 20 flowers. Corolla bell-shaped up to 3.7 cm across. **Description:** A small evergreen tree. **Bark** reddish brown, peeling off in small flakes. **Leaves** crowded at the ends of branches, 10-15 cm long, lance-shaped or elliptic oblong, pointed, leathery, rusty hairy or covered with small silvery scales beneath. **Flowers** large, very showy, commonly blood-red, rarely pink, very rarely almost white at higher altitudes; in about 10 cm wide clusters of about 20 flowers at the ends of the branches. Corolla bell-shaped, 2.5-3.7 cm across, lobes 5. Stamens 10 with white filaments. Capsules 2.5 cm long, ribbed, 10-chambered. **Seeds** minute. **Distribution:** All along the Himalaya up to Arunachal Pradesh, Meghalaya, Manipur, Myanmar and South East Tibet, at 1500-3600 m. Absent in the Kashmir valley proper. **Phenology: Flowers** February-May. **Fruits** in autumn and cold season. The fertilization of the flowers is carried out by insects and possibly by birds. Indian martens *Martes flavigula* hop from one cluster of flowers to another, thrusting their snouts to lick up the nectar. **Miscellaneous:** The flowers are offered in temples. They are sour to taste and are eaten, made into preserves or drunk as a sherbet in Kumaon. Wood is used for 'Khukri' handles and as fuel. The tender leaves are cooked as a vegetable. 'Guran' is the national flower of Nepal. Seedlings are raised by sowing seeds in March-April in pots filled with sand; the seeds should not be covered but sheltered from rain and sun and watered regularly. The seedlings are pricked in the second season. The easier method is to dig up seedlings from the forest and transfer them to the nursery.

78b. NILGIRI RHODODENDRON
R. arboreum ssp. *nilagiricum* (Zenker) Tagg.

Tamil *Billi.*

Under surface of leaves white to fawn with spongy tomentum; leaf apex rounded. Corolla crimson-blood red. Nilgiris above 1500 m. *R. arboreum* ssp. *zeylanicum*, with strongly concave leaves with blistered or puckered upper surface, lower surface with spongy fawn indumentum. Sri Lanka above 1500 m, gregarious.

79. MAHUA, BUTTER TREE
Madhuca longifolia (Koenig) MacBride var *latifolia* Chev.
(Family: Sapotaceae)

Syn. *Bassia latifolia* Roxb., *M. latifolia* (Roxb.) MacBride
M. indica J F Gmel.

Hindi, Gujarati, Bengali, Marathi *Mahua*; Tamil *Illupei*; Telugu *Ippa*; Malayalam *Poonam.*

Fig. 39

Field Identification: Leaves leathery, clustered at ends of branches 13-20 cm, elliptic-oblong on 2.5-3.7 cm long stalks. Flowers in clusters on 2.5-5 cm long stalks, 1.25 cm across. Corolla cream coloured, scented, fleshy and sugary, 8 or 9 lobed or 7-14 lobed falling off. Stamens 24-26, anthers on very short filaments. Style long and curved. Fruit 2.5-5 cm, juicy, seeds 1-4, brown and polished 2-3.3 cm long. **Description:** A large deciduous tree with a rounded crown. **Bark** grey, brown or blackish, with shallow wrinkles and cracks. **Leaves** leathery, clustered near the ends of branches, elliptic-oblong, 13-20 cm long on 2.5-3.7 cm long stalks which reveal drops of milky juice when cut. **Flowers** in clusters at or near ends of branches, 1.25 cm across, cream-coloured, scented, on 2.5-5 cm long woolly stalks. Calyx leathery, deeply 4-lobed. Corolla fleshy 1.5 cm, sweet, with 8 or 9 lobes, but sometimes 7 or up to 14, falling off early. Stamens 24-26, anthers on very short filaments. The ovary has a prominent long style. **Fruit:** a green juicy berry 2.5-5 cm long, containing 1-4 brown polished **seeds** 2-3.3 x 1.25-1.8 cm. **Distribution:** Himalayan foothills from the River Ravi to Kumaon and distributed southwards to Central India. **Phenology:** The **leaves** fall from February-April, new leaves about April or early May and are coppery red when young. **Flowers** February-April. Corollas fall soon after opening. **Fruit** ripens June to August. **Miscellaneous:** The scented, sweetish, fleshy flowers have an irresistible attraction for bears, deer and fruit and nectar-eating birds, such as mynas, bulbuls,

117

FIG. 39

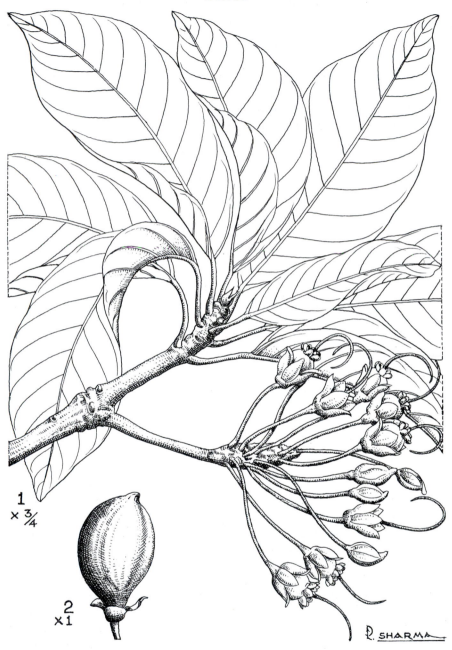

S.No. 79. *Madhuca longifolia* var. *latifolia*. 1) Shoot with leaves and flowers x $\frac{3}{4}$.
2) Fruit x 1.

parakeets, flowerpeckers and white-eyes. The birds are reported to get drunk when they peck fermented flowers. **Etymology:** *Madhuca* is derived from the Sanskrit *madhu* meaning honey, referring to the sugary flower petals, *longifolia* and *latifolia* (Latin) mean long leaves and broad leaves respectively. A spirit distilled from the flowers is a stimulant and appetizer. The oil from the seed is used to adulterate 'ghee' (and hence the popular name Butter Tree) and for the manufacture of soap and candles.

80. INDIAN MEDLAR TREE, BULLET WOOD
Mimusops elengi L.

(Family: Sapotaceae)

Hindi *Maulsiri*; Malayalam *Elengi*; Tamil *Mahila*; Burmese *Kaya*.

Fig. 40

Field Identification: Bark very dark, fissured lengthwise. Leaves shining 7-12 cm, oblong elliptic, pointed. Flowers 2 cm across, creamish, star-shaped, lobes 24. Fruit egg-shaped, orange-yellow, 2.5 cm long. Seed solitary, compressed, brown, shining 1.5-2.2 cm. **Description:** A large evergreen tree with a cylindric trunk and dense rounded crown. **Bark** very dark, fissured lengthwise and transversely. **Leaves** shiny, 7-11 x 3.5-5 cm, oblong, elliptic, abruptly pointed. **Flowers** fragrant, creamish, star-shaped, 1.2 cm across, corolla lobes 24, in 2 series. Stamens 8, staminodes 8. **Fruit** egg-shaped, orange yellow, about 2.5 cm long. **Seed** solitary, compressed, light brown, shining 1.5-2.2 cm. **Distribution:** Western Ghats from Maharashtra southwards, on the Eastern Ghats from North Circars southwards, Andamans, Myanmar and Sri Lanka. **Phenology: Flowers** February-April, July (Dehra Dun). **Fruit** ripens the following February-June. **Miscellaneous:** The fruit is edible. The oil from the flowers is used in perfumery, from the seed in cooking and as a luminant. The wood being very strong is used in bridges and house construction. **Etymology:** Medlar, according to the Oxford Dictionary, means "(Tree with) fruit like a small brown apple, eaten when decayed". The great Swedish botanist Linnaeus (1696-1759) selected its Malayalam name *elengi* as its specific name. The popular English name, Bullet wood was given because 'Maulsiri' is one of our strongest woods.

119

FIG. 40

S.No. 80. *Mimusops elengi*. 1) Branch with leaves and star-shaped flowers x 1.
2) Egg-shaped fruit x 2.

Fig. 41

S.No. 81. *Manilkara hexandra*. 1) Leaves notched at apex, and flower x 1. 2) An enlarged flower x 5. 3) Fruit x 1.

81. KHIRNI
Manilkara hexandra (Roxb.) Dubard

(Family: Sapotaceae)

Syn. *Mimusops hexandra* Roxb.

Hindi, Gujarati *Khirni*; Bengali *Krikhiyur*; Telugu *Pala*; Tamil *Palla*; Malayalam *Pala*; Marathi *Rayan*.

Fig. 41

Field Identification: Bark blackish-grey, blaze crimson, exuding milky juice. Leaves at ends of branchlets, 5-10 cm long, oblong, obtuse or notched at apex. Fruit olive-shaped, smooth, red, 1.25 cm long, with one (sometimes 2) reddish brown seeds. **Description:** A large tree with a spreading crown, stunted and reduced to a shrub in very dry regions. **Bark** blackish-grey, deeply furrowed vertically, blaze crimson, exuding drops of milky juice. **Leaves** crowded at ends of branchlets, oblong-obtuse, or notched at apex, 5-10 cm on 0.6-2 cm long stalks. **Flowers** pale yellow, 0.6 cm across, on 0.6 cm long stalks. Corolla lobes 18, 6 inner larger and 12 outer linear, stamens 6, staminodes 6. **Fruit** 1.25 cm long, olive-shaped, smooth and red. **Seeds** 1 (sometimes 2), shining reddish-brown. **Distribution:** Gujarat, Deccan, South India and dry regions of Sri Lanka. **Phenology:** Flowers November-February **Fruit** ripens April-July. **Miscellaneous:** The fruit is sweet and tasty and is collected in large quantities for food. The wood is very hard, heavy, tough and takes a fine polish. It is one of the important trees of Sri Lanka, used in panelling and furniture. It is planted as an avenue tree in U.P., Delhi and elsewhere in the Indogangetic plains.

82. BEEDI-LEAF TREE, COROMANDEL EBONY
Diospyros melanoxylon Roxb.

(Family: Ebenaceae)

Hindi, Marathi *Tendu*; Gujarati *Temru*; Tamil *Karai*; Telugu *Kumki*; Malayalam *Kari*.

Fig. 42

Field Identification: Bark blackish (characteristic of the genus). Male and female flowers on different trees. Calyx lobes 5. Male flowers small, in drooping cymes, stamens 10. Female flowers small, almost stalkless. Fruit round, 2.2 x 2 cm, yellow. Seeds 3-8, shining, embedded in the sweet pulp. **Description:** A medium-sized tree. **Bark** greyish-black, peeling off in rectangular scales.

FIG. 42

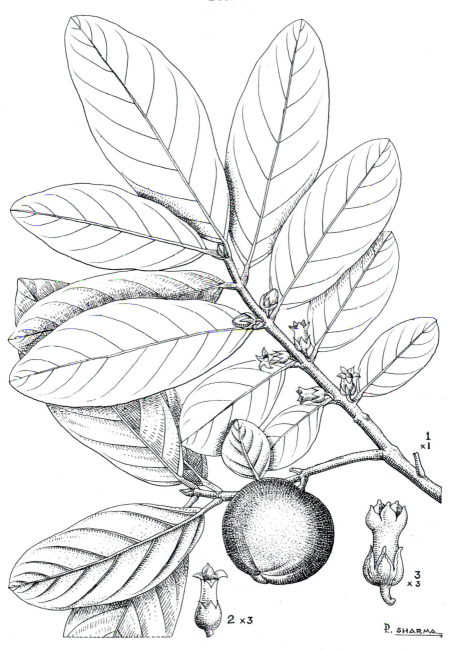

S.No. 82. ***Diospyros melanoxylon***. 1) Leaves, female flowers and fruit x 1. 2) Male flower x 3. 3) Female flower x 3. (From *Roxburgh T.* 46, *Brandis*, Indian Trees).

Leaves mostly sub-opposite, leathery, 6-14 x 5.6 cm, elliptic, egg-shaped, hairy beneath. Male **flowers** in drooping velvety cymes. Stamens 10, calyx lobes 5, pointed at the tip. Female flowers solitary, almost stalkless, calyx lobes heart-shaped, broadly   2-winged, corolla lobes yellow. **Fruit** globose, 2.2 x 2 cm, smooth, yellowish when ripe. **Seeds** 3-8, 1.2-2 cm, brown, shining, embedded in sweet, yellow, edible pulp. **Distribution:** Peninsular India, Central India and in Sri Lanka in dry deciduous forests. **Phenology:** Leafless briefly in the hot season, never completely leafless. **Flowers** April-June. **Fruit** ripens in April-June of the following year. **Miscellaneous:** The fruits are devoured by fruit-bats, monkeys, birds, particularly hornbills, which are seen among the trees at the time when fruit ripens. The seeds are scattered by them. The leaves are in great demand for rolling 'beedis' and are collected in government forests. The heartwood which is black, streaked with purple or brown, is used in cabinets, inlay work and brush handles, etc.

83. EBONY, TRUE EBONY
Diospyros ebenum Koen.

(Family: Ebenaceae)

Hindi *Ebans*, *Abnus*; Tamil *Karunkali*; Telugu *Tuki*; Sri Lanka *Kaluwaru*.

Fig. 43

Field Identification: Small-medium sized in India, large tree in Sri Lanka. Bark dark grey. Leaves oblong-lance-shaped 5-10 cm, leathery. Flowers unisexual and bisexual on the same tree. Flowers tetramerous i.e. floral parts in fours or multiples of four. Fruit round 1.25-2 cm across, seeds 8 black. **Description:** Small or medium-sized tree in India, large in Sri Lanka, evergreen, with a dense crown; young shoots softly hairy. **Bark** dark grey. **Leaves** oblong-lance-shaped, 5-10 cm long, thinly leathery. **Flowers** unisexual and bisexual on the same tree; male flowers 3-6, in shortly stalked clusters, corolla greenish yellow, stamens 16, female flowers solitary, calyx densely rusty hairy within. **Fruit** round, 1.25-2 cm across, seated in an enlarged woody, cup-shaped calyx. **Seeds** 8 black. The flowers are tetramerous i.e. the floral parts, calyx, corolla, stamens, etc. are in fours or in multiples of four. **Distribution:** Hills of the Deccan, up to Cuddapah and Kurnool (Andhra Pradesh), Kerala and Sri Lanka. Less frequent in the Deccan, but more frequent in Sri Lanka chiefly in dry evergreen forests. **Phenology:** In Sri Lanka, **flowers** about March, but the flowering season is irregular. **Fruit** ripens September-October. **Miscellaneous:** It is the true ebony of commerce and is of great commercial importance in Sri Lanka. The heartwood is jet-black. Used in carving, cabinet work, piano keys and as chop-sticks in China.

Fig. 43

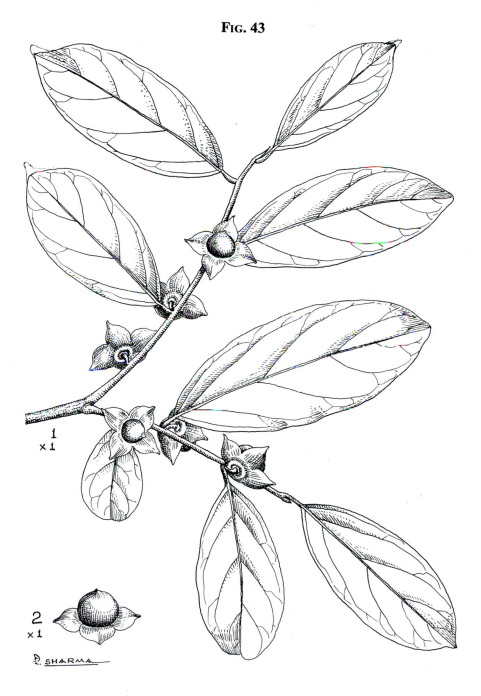

1
× 1

2
× 1

D. SHARMA

S.No. 83. *Diospyros ebenum*. Leaves and young fruits x 1. 2) Fruit x 1. (From *Gamble* 15741, *Mooney* 2874).

84. ANDAMAN MARBLE WOOD, ZEBRA WOOD
Diospyros marmorata Parker

(Family: Ebenaceae)

Andamans *Kala Takir.*

Fig. 44

Field Identification: Leaves alternate, 7-11 x 2.5-6 cm, ovate-elliptic, pointed, base rounded; lateral nerves *c.* 9 pairs. Male flowers in pairs of 3-4. Stamens 12, 3 at the base, 6 higher up, and 3 at the base of the corolla lobes, anthers 2 mm, pointed. Fruit 3 cm, round, 6-celled. Seeds 3, endosperm mottled. **Description:** An uncommon tree endemic to the Andamans. **Leaves** alternate, 7-11 cm long by 2.5-6 cm broad, ovate-elliptic, pointed, base rounded, leathery, lateral nerves *c.* 9 pairs, inconspicuous, leaf stalk about 5 mm long. **Flowers:** Male flowers axillary in panicles of 3-4, flower stalk 4-5 mm long, calyx 5-6 mm, 3-toothed, corolla 12-14 mm long, minutely hairy, 3-lobed, overlapping to the right. Stamens 12, 3 inserted at the base of the corolla tube, 6 inserted slightly irregularly higher up and 3 inserted half way up the tube. Anthers minutely hairy, 2 mm long, pointed. **Fruit** 3 cm in diameter, round, 6-celled. **Seeds** 3, endosperm mottled. **Distribution:** Long Island, Andamans. Full distribution data lacking for the Andamans and Nicobars. **Phenology:** Data lacking. **Miscellaneous:** It is one of the most decorative timbers of the world. The wood is figured with jet-black stripes interspersed with light brown bands and takes a beautiful polish. Only limited supplies are now available. Being endemic it is on the Endangered List of Flora. Average size of logs 3-5.4 x 0.6-0.9 m. Used for table tops, cabinets, brush backs, etc. The Andaman marblewood was believed to be produced by *Diospyros kurzii* Hiern, due to botanical specimens and the corresponding timber specimens becoming mixed up. It is called Marble wood because of the marble-like look of the polished timber, also called Zebra Wood because of the jet-black stripes in the wood. Endemic to the Andamans and earlier referred as *D. oocarpa,* of the Western Ghats, also as *D. kurzii* Hiern. R.N. Parker, Forest Botanist, Forestry Institute, Dehra Dun sorted out the taxonomic confusion in a paper in the *Indian Forester* LVII: 209-211, 1931 and established a new species *D. marmorata* Parker. He obtained botanical specimens from the Andamans and found out the differences between the trees from Western Ghats and the Andamans. He obtained good fruiting material also but failed to get female flowers which apparently have not been studied so far.

Fig. 44

S.No. 84. ***Diospyros marmorata***. 10 Shoot with leaves, flowers x 1. 2) Flower x 2.
3) Opened flower showing 12 stamens x 3. (From *R.N. Parker*). 4) Fruit x 1.

85. CORAL JASMINE, TREE OF SORROW
Nyctanthes arbor-tristis L.

(Family: Oleaceae)

Hindi *Harsinghar*; Marathi, Gujarati *Parijat*; Malayalam *Parijatukam*; Telugu *Parjatamu*; Burmese *Shephalika*.

Field Identification: Shrub to small tree, with drooping quadrangular branchlets and rough opposite leaves 7.5-10 x 6-7 cm. Flowers fragrant, creamish, corolla salver-shaped with an orange tube. Capsule 1.2 cm. **Description:** A large deciduous shrub or small tree with drooping branches and quadrangular branchlets. **Leaves** opposite, ovate, rough, with stiff white hairs, entire or coarsely toothed, 7.5-10 x 6-7 cm, pointed, lateral nerves 4 pairs. **Flowers** fragrant, stalkless, in clusters of 3-7 forming terminal cymes, corolla salver-shaped, 5-8 lobed, creamish, corolla tube 1.2 cm, orange coloured, stamens 2, anthers very short stalked. **Capsule** flat, roundish, 1.25 cm across, 2-seeded. **Seed** erect, round and flattened. **Distribution:** Sub-Himalayan tracts, Central India and Sri Lanka. **Phenology:** Flowers August-October. **Fruits** November-February. Flowers open towards the evening and drop the next morning. **Miscellaneous:** It is cultivated in gardens and temples for its fragrant ivory white flowers. The fallen flowers look decorative carpeting the lawns or floating in the garden ponds. They are gathered for Hindu ceremonies and made into garlands in Sri Lanka, also for dyeing silks. The leaves, which are rough like sand paper serve for polishing wood and utensils. Useful for afforesting denuded Himalayan slopes where it spreads quickly. It reproduces readily from seed. **Etymology:** The flowers open during the night and fall off the tree in the early morning, hence probably the specific name *arbor-tristis*. Owing to its drooping foliage and flowers it is called the "tree of sorrow", and coral jasmine for its coral-like flowers.

86. INDIAN OLIVE
Olea ferruginea Royle

(Family: Oleaceae)

Hindi *Kahu*; Pushtu, Pakistan *Zaitun*.

Fig. 45

Field Identification: Small tree of the drier Western Himalaya. Leaves 5-10 cm long, leathery, pointed, reddish beneath. Flowers small, whitish, in 2-4 cm long axillary clusters. Fruit black when ripe, 8 mm long, not pleasant to eat. **Description:** A small evergreen tree, with grey bark which strips off when old. **Leaves** dark green, leathery, oblong-lance-shaped, with a rigid pointed tip, 5-10 cm long, with

FIG. 45

S.No. 86. **Olea ferruginea**. 1) Leaves, inflorescence x 1. 2) Fruits x 1. (From *Bisram* 53416 and *Wingar* s.n. Chitral).

a dense film of minute scales beneath which turn reddish brown when old, leaf margin slightly recurved. **Flowers** whitish in 2-4 cm long clusters; corolla tube short, deeply divided. **Fruit** 8 mm, egg-shaped, black when ripe, one seeded, pulp scanty, oily. **Distribution:** Western Himalaya from Yamuna westwards to Pakistan up to 2400 m in the inner dry valleys. **Phenology: Flowers** April-May. **Fruits** August-November. **Miscellaneous:** It makes a suitable stock on which to graft the cultivated Olive, *O. europaea*, to which it is related. This has been done on a large scale in Himachal Pradesh and Western U.P. with Italian collaboration. **Etymology:** The specific name *ferruginea* in Latin means rusty, pointing to the leaves which are rusty red beneath.

87. SCHOLAR TREE, DEVIL'S TREE
Alstonia scholaris R. Br.

(Family: Apocynaceae)

Hindi *Shaitanki jhur*; Bengali *Chattim*; Tamil, Malayalam *Pala*; Sanskrit *Saptaparni*.

Fig. 46

Field Identification: Large buttressed tree, branches whorled, exuding milky latex when cut. Leaves whorled, 7.5-20 x 2.5-3.7 cm, oblong-lance-shaped, round or notched above, lateral nerves parallel, uniting near the margin into a vein. Flowers small, greenish-white, in flat topped clusters. Fruit made up of two slender, hanging follicles over 30 cm long. **Description:** A large evergreen tree often fluted or buttressed; branches whorled. **Bark** grey, yellow inside, exuding milky latex when cut. **Leaves** 4-7 whorled, 7.5-20 x 2.3-3.7 cm, oblong lance-shaped, obtuse or notched at the tip, stalkless or short stalked, bright green, shining above, lateral nerves numerous, parallel, uniting close to the margin in a distinct vein. **Flowers** small, greenish white, in compact, flat-topped clusters. Calyx small, 5-lobed; **corolla** 0.7-1.2 cm across, lobes rounded, twisted in buds. **Fruit** of two long slender follicles over 30 cm long and 0.55 cm wide, hanging in clusters. **Seeds** linear-oblong, 0.5 cm long with tufts of hair at each end. **Distribution:** Sub-Himalayan tracts from Yamuna eastwards up to 800 m. Scattered through the greater part of India and Myanmar in moist deciduous forests. **Phenology: Flowers** December-March. **Fruits** May-July. **Miscellaneous:** It is a popular avenue and park-tree, seen in avenues in New Delhi and other cities. The soft white wood is used for tea chests, packing cases and match splints. **Etymology:** It gets its specific name *scholaris* from the fact that in bygone days the wood was used for making school slates. The tree is shunned by animals because of its poisonous nature and hence the name Devil's tree.

FIG. 46

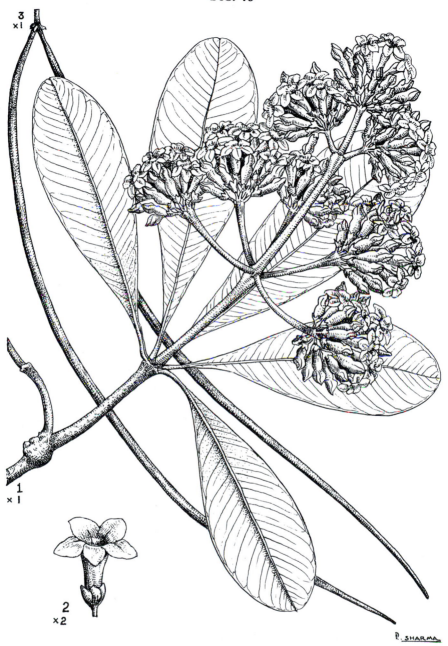

S.No. 87. *Alstonia scholaris*. 1) leaves in whorls, inflorescence x 1. 2) Flower x 2.
3) Two slender follicles x 1. (From *Sis Ram* s.n. and *Raizada* s.n.)

88. STRYCHNINE TREE, NUX-VOMICA TREE, POISON NUT TREE, SNAKE WOOD

Strychnos nux-vomica L.

(Family: Apocynaceae)

Plate 9

Hindi *Kagra*; Marathi *Kar*; Tamil *Yetti*.

Field Identification: Small to large spiny tree. Leaves opposite, broadly elliptic, 5-12.5 cm, 5-nerved. Flowers greenish white, 1.2 cm long, corolla 5-lobed. Fruit hard globose, of the size and colour of a small orange, filled with bitter white pulp and nearly circular, silvery seeds with silky hairs. **Description:** A small-moderate sized or large handsome evergreen or deciduous tree. **Bark** yellowish-grey to blackish-grey, thin, smooth, covered with minute warts. Branches opposite, often converted into strong, woody spines. **Leaves** opposite, broadly elliptic, 5-12.5 cm long, shining, 5-nerved from the base, the 3 central nerves prominent, leaf stalk 0.6-1.2 m. **Flowers** greenish white, 1.2 cm long, corolla 5-lobed, corolla tube 0.75 cm long. **Fruit** hard, globose, of the size and colour of a small orange. **Seeds** nearly circular, 2 cm across, silvery and shining, with silky hair radiating from the centre of the seeds, immersed in a bitter white pulp. **Distribution:** Bengal, South India. Common in the moist monsoon forests of the Western Ghats, also along the sea coast in evergreen scrub forests. **Phenology:** Evergreen in moist forests. In dry forests, leaves are shed for a short time in the hot season. **Flowers** March-May. **Fruit** ripens at various times throughout the year but mostly from December-June. **Miscellaneous:** The seeds are poisonous, the pulp is eaten by langurs, flying foxes, hornbills, parakeets and other birds. The seeds are probably ejected or passed out undigested. It is called the strychnine tree or poison nut tree because its seeds are the source of the poisonous alkaloid strychnine. The seeds are the nux-vomica of commerce, which contains the valuable alkaloid strychnine, also used in the distillation of country spirits to make them more potent. The wood is not attacked by termites and is used in ploughs, axe handles, etc.

89. INDIAN CORK TREE

Millingtonia hortensis Linn.

(Family: Bignoniaceae)

Hindi *Akas-nim*; Burmese *Ay-Ka-Yit*.

Fig. 47

Field Identification: Tall tree, bark dark grey, corky. Leaves large, divided into leaflets. Flowers fragrant, white in hanging masses. Corolla 2.5 cm across with a

FIG. 47

1
× ½

2 × 2

D. SHARMA

S.No. 89. *Millingtonia hortensis*. 1) Leaves and flowers x ½. 2) Corolla cut open showing stamens x 2.

slender 5-7.5 cm long tube. Fruit a linear 30 cm long x 2 cm wide capsule with winged seeds 2.5 cm across. **Description:** A tall, handsome, evergreen tree with dark grey, corky bark, 2.5 cm thick with a yellowish cut. Leaves large, much divided into 2.5-5 cm long leaflets. **Flowers** fragrant, white, in hanging masses. Corolla salver-shaped, 2.5 cm across, with a long slender tube 5-7.5 cm long. Fertile stamens 4, without a sterile one, one of the anther cells spurred. **Fruit** a linear capsule 30 cm long x 2 cm broad, splitting into 2 valves when ripe. **Seeds** in many rows, 2.5 cm across, surrounded by a gland dotted delicate wing. **Distribution:** Myanmar, tropical forests from Martaban southwards to Tenasserim, rare, also in Ava. Widely cultivated in India and Pakistan. **Phenology: Flowers** October-November. **Fruits** end of the hot season. **Miscellaneous:** A fast growing ornamental tree, suitable for parks. Being brittle, shallow rooted and prone to be uprooted in hurricanes, it is not suitable as an avenue tree or for planting near buildings. Bark used as an inferior substitute for cork. **Etymology:** *Millingtonia* is named in honour of T. Millington, an English botanist of the 18th century; *hortensis*, in Latin means pertaining to a garden. The tree was known only under cultivation, its natural habitat remaining uncertain.

90. WAVY-LEAVED TECOMELLA
Tecomella undulata (Smith) Seemann

(Family: Bignoniaceae)

Hindi *Rugtrora*, Punjabi *Lahura*; Rajasthan *Roira*; Waziri *Ribdavan*; Baluchistan *Rori*; Sind *Lohero*.

Field Identification: Small handsome tree of the semi-arid parts of Rajasthan, etc. with a round topped crown and drooping branches. Leaves wavy, narrowly oblong, 5-12 x 3.5 cm. Flowers orange, trumpet-shaped, corolla 5 cm long, 5-lobed, in clusters of 5-10. Capsule 15-17 cm, slightly curved, sharply pointed, with winged seeds. **Description:** A large evergreen shrub to a small handsome tree, topped with a fine rounded crown, the extremities of the branches drooping. **Leaves** greyish-green, simple, 5-12.5 x 1.2-3.2 cm, narrowly oblong, with a wavy margin, leaf stalk 2 cm. **Flowers** large, from pale yellow to deep orange, odourless, in clusters of 5-10 at the ends of the smaller lateral branches. Stalk 0.6-1.2 cm long; calyx bell-shaped, nearly 2 cm, 5-lobed. Stamens 4, protruding out. Corolla trumpet-shaped, about 5 cm long, 5-lobed. **Capsule** 15-17.5 cm long, slightly curved, sharply pointed at the tip. **Seed** winged, 2.5 cm x 0.37 cm (including the wing); wing thin, very narrow, rounded at the top and absent at the base of the seed. **Distribution:** Pakistan, Punjab, Rajasthan, extending eastwards to Yamuna, Gujarat and Western peninsular India up to 800 m. **Phenology: Flowers** March-

April. **Fruits** May-July. **Miscellaneous:** Raised from seed and cuttings. It needs to be popularised as it is a very beautiful tree and because of its very small size, ideal for today's small gardens. It is cultivated in parks and gardens. Useful for afforesting dry tracts and is resistant to fire. The wood is prized for carvings and furniture and goes by the name of Rajasthan Teak.

91. GAMARI
Gmelina arborea Roxb.
(Family: Verbenaceae)

Hindi *Gamhar*; Bengali *Gambhar*; Assamese *Gomari* Telugu *Gummadi*; Tamil *Kumadi*; Malayalam *Kumbil*; Marathi, Gujarati *Shewan*.

Fig. 48

Field Identification: Leaves opposite, heart-shaped, 8-15 cm long, with star-shaped hairs beneath. Flowers yellow, in 15-30 cm long panicles, corolla 3.7 cm long, 5-lobed, lobes recurved. Drupe pear-shaped, 2.5 cm long, orange-yellow when ripe with sweet edible pulp. Stone egg-shaped, pointed at one end, 1.5-2.5 cm, 2-seeded. **Description:** A moderate-sized to large handsome flowering tree with a straight trunk. **Bark** light grey, smooth, corky, yellow inside. **Leaves** deciduous, opposite, heart-shaped, 8-15 cm long, with a pointed tip, with star-shaped hairs beneath, leaf stalk 5-7.5 cm long with shining glands on top. **Flowers** yellow, produced in profusion in spring when leafless, in 15-30 cm long terminal panicles; calyx bell-shaped. Corolla yellow tinged with brown, 3.7 cm long, lobes 5, 2-lipped, lobes spreading, recurved. **Drupe** egg-shaped or pear-shaped, 2.5 cm long, orange yellow when ripe. Stone egg-shaped, bony, pointed at one end, 1.5-2 cm, two-seeded. **Distribution:** Widespread in deciduous forests in India, Nepal, Bangladesh, Myanmar and Sri Lanka, up to 1300 m, attaining large size in the Eastern sub-Himalayan tracts. **Phenology: Leaves** fall in January-February, new leaves in March-April. **Flowers** March-April, corollas fall quickly and cover the forest floor. **Fruits** April-July. **Miscellaneous:** Young shoots are browsed by deer. The wood is strong and does not warp and therefore is in demand for musical instruments and dugouts. It is a fast growing tree, and therefore popular in the tropics for its valuable timber. It is also an ornamental tree. The fruit is eaten by the Bhils and other tribes.

FIG. 48

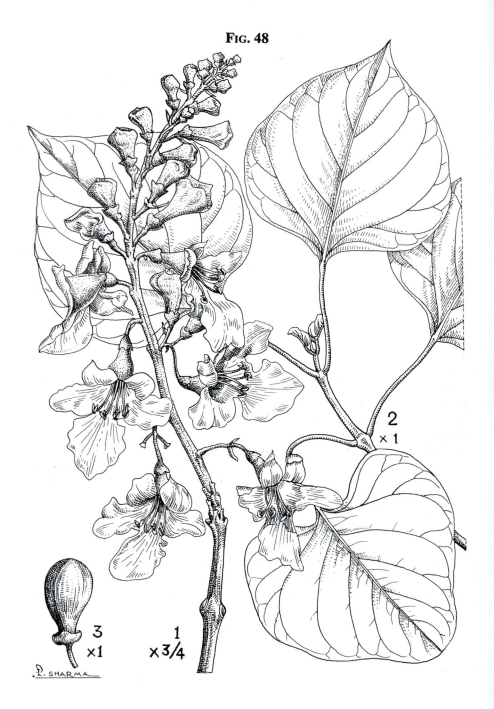

S.No. 91. *Gmelina arborea*. 1) Inflorescence x $\frac{3}{4}$. 2) Leaves x 1. 3) Fruit x 1.

92. TEAK
Tectona grandis Linn.
(Family: Verbenaceae)

Hindi *Sagun, Sagwan*; Gujarati, Marathi *Sag*; Bengali *Segun*;
Telugu *Teeku*; Tamil *Tekkumaram*; Malayalam *Thekku*.

Fig. 49

Field Identification: Branchlets quadrangular and channelled. Leaves very large, 30-60 cm x 20-30 cm, opposite, rough, sandpaper-like above, undersurface with star-shaped hairs. When held against the light, small red dots are visible on the leaf surface which turn black at maturity. Young leaves when rubbed between the hands stain the palms red - a test for genuine teak trees since early times. Flowers small white in metre-long panicles. Fruit bladder-like. **Description:** A large deciduous tree, up to 45 m or more in height under favourable conditions. **Bark** fibrous, light brown or grey, peeling off in thin flakes. Branchlets quadrangular and channelled. **Leaves** opposite, broadly elliptical or obovate, 30-60 cm x 20-30 cm, rough sandpaper-like above, greyish hairy beneath, hairs star-shaped. They possess minute red glandular dots which turn black at maturity. **Flowers** small, white, in conspicuous 1 m long erect panicles. **Fruit** hard, bony, 1-1.5 cm in diameter, 4-celled, enveloped by a bladder-like calyx. **Seeds** 1-3, rarely four, marble-white, egg-shaped, 4-8 mm long. **Distribution:** Indigenous to Western Peninsula, Central India, Bihar and Myanmar. The great belt of teak forests in central India can be seen while travelling by train. Nilambur in Kerala is famous for its teak forests planted in 1844. Here it attains the largest dimensions, up to 56.5 m in height. The maximum height recorded is 56.5 m in Malabar, the maximum girth, 8 m in Myanmar. **Phenology: Leafing, flowering and fruiting** April to June. Leafless throughout the greater part of the hot season. **Flowers** June to August or September. **Fruits** November to January. **Miscellaneous:** The timber is world-famous and its uses are well known. The heartwood is golden yellow when freshly cut, later turning brown. The hard knots which develop on trunks are prized for making tobacco pipes. It is widely used for making decks of ships and rightly called ship tree. On the basis of origin it is distinguished into two classes, Central Indian teak and South Indian teak. In the former, the wood has a darker shade with wavy streaks and in the latter it is light coloured and more straight grained. In the Anaimalais, it is reported to have attained an age over 500 years. A tree cut down in Kakanankote forests of Mysore showed 680 annual rings! **Etymology:** The generic name comes from the Malayalam '*tekku*'.

Fig. 49

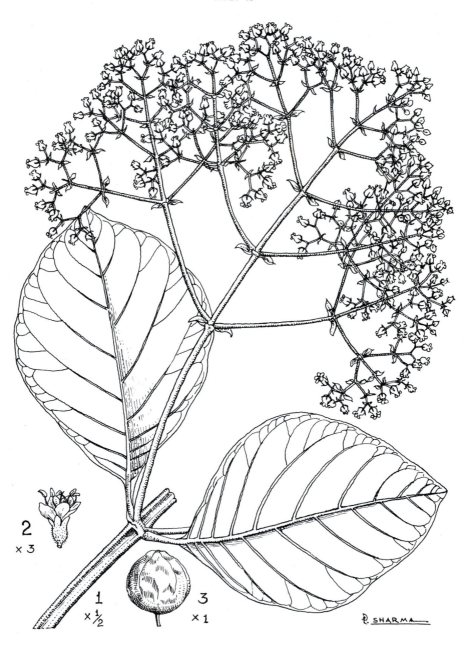

S.No. 92. ***Tectona grandis***. 1) Leaves, inflorescence x $\frac{1}{2}$. 2) Flower x 3. 3) Fruit enveloped in bladder-like calyx x 1.

Fig. 50

S.No. 93. *Cinnamomum tamala*. 1) Triple-nerved leaves and inflorescence x 1.
2) Fruits x 1. (From *Smythies* s.n. and *Subramanium* s.n.).

93. CINNAMON
Cinnamomum tamala (Spreng.) Nees & Eberm.

(Family: Lauraceae)

Sanskrit *Tejpatra*; Hindi, Bengali *Tejpat*; Gujarati *Tamalpatra*;
Tamil *Taalishappatri*; Telugu *Talispatri*; Burmese *Thitchabo*.

Fig. 50

Field Identification: Bark thin, wrinkled, blaze reddish brown, very aromatic.
Leaves alternate and opposite on the same twig, 6-15 x 3.7-7 cm, elliptic, pointed.
3-nerved, leaf stalk 0.7-1.2 cm. Flowers yellowish white, 0.6 cm across, perianth
of 6 unequal segments, silky hairy. Perfect stamens 9, the outer 6 without glands,
the inner 3 with 2 glands, innermost of 3 short staminodes. Drupe 1.25 cm long,
egg-shaped, black when ripe. **Description:** A medium sized evergreen tree. **Bark**
thin, dark-brown, wrinkled, blaze reddish brown, very aromatic. Buds and leaves
silky hairy. **Leaves** both alternate and opposite on the same twig, 6-15 x 3.7-7 cm,
egg-shaped-oblong, pointed, 3-nerved, leaf stalk 0.7-1.2 cm, shining above, pink
when young. **Flowers** 0.6 cm across, yellowish white, perianth of 6 unequal
segments, silky hairy. Perfect stamens 9, the 6 outer without glands, the inner 3
with 2 glands at the base, innermost of 3 short staminodes. **Drupe** 1.25 cm long,
egg-shaped, black when ripe. **Distribution:** Sub-Himalayan tracts from Yamuna
eastwards. Upper Assam, Khasi Hills, Tripura, Bangladesh and Upper Myanmar.
From 800-1600 m chiefly in damp ravines. **Phenology:** Young leaves pink, April-
May. **Flowers** February-May. **Fruit** ripens June-October, remaining for long on
the tree. **Miscellaneous:** The leaves known as 'tejpat' are aromatic and used as
food flavouring. The bark is used as a substitute for the true cinnamon *C. verum*.

94. TRUE CINNAMON TREE
Cinnamomum verum J.S. Presl.

(Family: Lauraceae).

Hindi *Shudh Dalchini*; Kannada *Dalchini*; Gujarati *Taj*;
Burmese *Hmanthin*.

Fig. 51

Field Identification: Bark thick, inner bark reddish. Leaves opposite and alternate,
very aromatic, elliptic, 7.5-20 x 3.7-7.4 cm, leathery, dark green, 3-5 nerved.
Flowers in loose panicles 2-3 together, perianth 0.6 cm, lobes broad, egg-shaped.
Fruit egg-shaped, 1.2 cm, minutely pointed, dark purple, seated on an enlarged
perianth. **Description:** A medium-sized to large evergreen tree. **Bark** 0.6 cm thick,

140

FIG. 51

S.No. 94. ***Cinnamomum verum***. 1) Leaves, inflorescence x 1. 2) Flower x 3.
3) Fruit x 1. A) Bark.

smooth, pale coloured, rough in old trees, inner bark reddish; branchlets smooth, compressed and grooved. **Leaves** opposite or alternate, very aromatic when crushed, elliptic, 7.5-20 x 3.7-7.4 cm, pointed, leathery, dark green above, nerves 3-5, sub-basal, strong beneath, young leaves pink. **Flowers** in loose panicles 2-3 together; flower stalks 0.2-0.6 cm, hairy. Perianth 0.6 cm long, lobes broad, egg-shaped. **Fruit** egg-shaped, 1.2 cm long, minutely pointed, dark purple, seated on the much enlarged, wrinkled, perianth tube. **Seed** one. **Distribution:** Western Ghats, very common in North Kanara. Moist low areas in Sri Lanka and Myanmar. **Phenology: Flowers** November-February. **Fruits** June-July. **Miscellaneous:** Fruit greedily eaten by pigeons, crows and other birds, and thus seeds get dispersed. True 'dalchini' or cinnamon spice is obtained from the inner bark of shoots in the form of rolled quills. Cinnamon from Kanara is not as good as that from Sri Lanka. The leaves are used for flavouring food. The oil extracted from the bark is golden yellow, sweet, aromatic and is used in perfumes. **Etymology:** The specific name *zeylanicum* in Latin refers to Ceylon (now Sri Lanka) where the best quality occurs. The oldest correct name for *C. zeylanicum* is *C. verum.*

95. AGARWOOD TREE
Aquilaria malaccensis Lamk., *A. agallocha* Roxb.
(Family: Thymeliaceae)

Hindi, Bengali *Agar*; Assamese *Sasi*; Burmese *Akyaw*.

Fig. 52

Field Identification: Leaves alternate, leathery, 5-8.7 cm, lance-shaped, pointed with faint parallel nerves. Flowers white in flat-topped clusters on thin 0.6 cm long stalks. Perianth bell-shaped, 0.6 cm long. Stamens 10, filaments red at apex. Fruit 2.5-3.7 cm egg-shaped, velvety. Wood scented. **Description:** A tall evergreen tree. Young shoots silky. **Leaves** alternate, thinly leathery, lance-shaped, pointed, 5-8.7 cm long, with parallel intermediate nerves. **Flowers** white, in flat-topped silky clusters, flower stalks slender, 0.6 cm long. Perianth 0.6 cm long, bell-shaped, densely silky and hairy within. Stamens 10, filaments red at apex. **Fruit** velvety, 2.5-3.7 cm, rarely up to 5 cm, egg-shaped, Wood scented due to the presence of a resinous substance. **Distribution:** East Himalayan low hills, Assam, Bangladesh and Myanmar, mostly in Martaban hills. **Phenology: Flowers** April-August. **Fruiting** winter, but generally in tropical, subtropical areas, flowering and fruiting goes on all the year round. **Miscellaneous:** The wood is scented and used as incense and commands a very high price. Diseased trees become infiltrated with a resinous substance. It is the diseased wood that is valuable, for the hard dark coloured masses are caused by a fungus, resulting in the eaglewood or agarwood of commerce. The light coloured wood in which the resin is embedded is distilled

142

FIG. 52

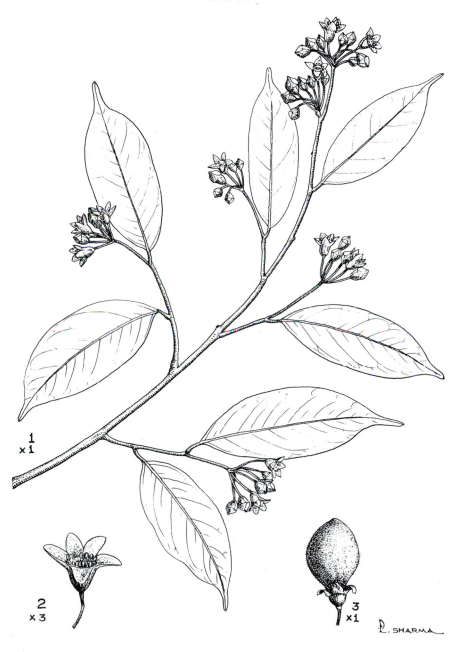

$\frac{1}{\times 1}$

$\frac{2}{\times 3}$

$\frac{3}{\times 1}$

P. SHARMA

S.No. 95. *Aquilaria malaccensis*. 1) Leaves, inflorescence x 1. 2) Flower x 3.
3) Fruit x 1. (From *Purkayastha* s.n. and *Gamble* 5887).

into an oil called 'agar-attar' used in perfumes. The non-infected wood is used by the Karens of Myanmar for bows. The bast of the bark when treated resembles parchment and was used by the kings of Assam to write upon in ancient times.

96. SANDALWOOD TREE
Santalum album L.

(Family: Santalaceae)

Fig. 53

Hindi *Sandal*; Punjabi, Gujarati, Marathi *Chandan*; Bengali *Chandano*; Kannada *Gandha*; Tamil *Sandanam*; Malayalam *Chandanam*; Burmese *Santagu*.

Field Identification: Small evergreen tree with drooping branches. Bark dark grey. Leaves opposite, egg-shaped, shining, 3.7-6.2 cm long. Flowers small, purple brown, perianth bell-shaped, 4-lobed, fruit purple black, 0.6-1.2 cm across. **Description:** A small evergreen tree parasitic on roots of other plants, with slender drooping branchlets. **Bark** dark grey. **Leaves** opposite, sometimes alternate, egg-shaped or elliptic lance-shaped, shining, 3.7-6.2 x 1.5-3.2 cm on 0.6-1.2 cm long stalks. **Flowers** small, odourless, in loosely branched clusters, purple-brown, perianth bell-shaped, with 4 triangular lobes. Stamens 4. **Drupe** globose, 0.6-1.2 cm, purple black. **Seeds** bony, rough, single. **Distribution:** Probably indigenous to south-west India from Nasik and Northern Circars southwards. Endemic to Timor Islands. Some botanists consider that it was introduced to India from Timor in ancient times. There is evidence that it has been in India for 23 centuries as there are references to it or possibly to red sanders in the Pali *Milinda-panha* (150 B.C.) and the *Mahabharata*. This is an unsolved problem awaiting fossil findings and discovery of pollen grains deposited in sediments in the ancient past. India is now the main source of sandalwood. **Phenology:** True evergreen, flush of new leaves in May. **Flowers** February-April. **Fruit** ripens May-June. **Miscellaneous:** Birds relish the fruit and are important dispersal agents. Sandal trees are found well inside forests in South India, reinforcing the theory that Sandal may not be indigenous to India but spread deep inside the forests in South India through the agency of birds after the initial introduction by man in ancient times on the outskirts of forests and near villages. *Host:* The sandal tree is a parasite on the roots of some trees; most members of the Mango family are fatal to sandal. The strychnine tree (*Strychnos nux-vomica*), native to the Western Ghats, is a reasonably good host. *Spike disease:* Sandal tree is of interest because it is prone to the widely known disease called Spike disease. This is caused by a virus but now considered a mycoplasma disease not known outside India, which causes damage worth several

Fig. 53

S.No. 96. **Santalum album**. 1) Leaves, inflorescence x 1. 2) Open flower (enlarged). 3) Fruit x 1.

lakhs of rupees every year. *Uses:* It is famous for its strongly scented yellow heartwood used in carving. The heartwood also yields a fragrant oil used in perfumery and soaps. Sandal was an important source of revenue to Mysore where it was declared a Royal Tree, which meant that only the Government could commercially exploit trees growing within the State borders.

97. AMLA, EMBLIC MYROBALAN
Embilica officinalis Gaertn.

(Family: Euphorbiaceae)

Syn. *Phyllanthus emblica* Linn.

Hindi, Gujarati *Amla*; Marathi *Avla*; Tamil, Malayalam *Nelli*; Kannada *Amalaka*; Burmese *Zibya*.

Fig. 54

Field Identification: Small to medium sized tree. Leaves small, 1.2 cm long, narrowly linear, closely borne on branchlets, feather-like and graceful. Flowers small, greenish-yellow, clustered along the branches. Fruit globose, greenish-yellow, 2 cm across, in clusters, fleshy, astringent. **Description:** A small to medium-sized deciduous tree, sometimes with a gnarled trunk. **Bark** greyish brown, peeling off in scroll-like patches, exposing the yellowish buff under-layer. Its feathery light-green **foliage** makes it look very graceful. **Leaves** small, narrowly linear, 1.2 cm x 0.3 cm, closely borne on deciduous branchlets. **Flowers** small, greenish-yellow, densely clustered along the branches, male flowers numerous on short slender stalks, female flowers few. **Fruit** looks like goose berries clustered along the branches, globose, fleshy, up to 2 cm across, with 6 vertical, pointed furrows, yellow when ripe. **Seeds** 4-6, dark brown, smooth. **Distribution:** Common throughout the Subcontinent up to 1300 m, except the arid regions, also in Myanmar and Sri Lanka. Widespread in Uttar Pradesh. **Phenology: Leaves** are shed during November-December. Leafless February or March. **Flowers** March-May when they are visited by swarms of bees. **Fruit** ripens from November-February. **Miscellaneous:** The fruit is probably the richest known natural source of vitamin C. It is often eaten as a thirst quencher and made into pickles and preserves. Fruit, bark and leaves are used in tanning.

146

FIG. 54

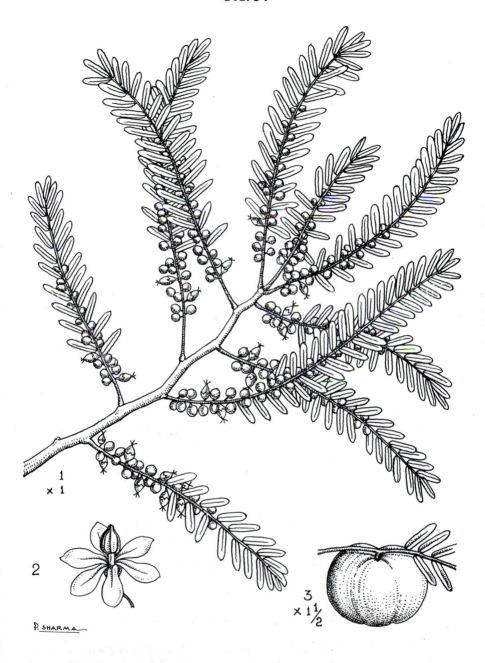

S.No. 97. **Embilica officinalis**. 1) Flowering shoot with leaves and young fruit x 1.
2) Flower enlarged. 3) Fruit x 1½.

98. BLINDING TREE
Excoecaria agallocha L.

(Family: Euphorbiaceae)

Bengali *Geor*; Marathi *Geva*; Tamil *Tilai*; Malayalam *Komatfi*;
Burmese *Tayaw*.

Fig. 55

Field Identification: Small mangrove tree with vertical branches, scars of fallen leaves visible. Leaves alternate, bright green, shining, elliptic, pointed with unbroken margins or toothed, 5-10 x 3.7-5 cm. Flowers minute, yellowish-green, male in 3.7-5 cm long spikes, male flowers stalkless; female flowers in shorter spikes, flowers stalked. Fruit a 3-lobed capsule, 0.7-1.2 cm. **Description:** A small mangrove tree with an inclined trunk, with vertical branches sticking out. **Bark** grey, smooth, with red-brown lenticels forming diagonal or vertical stripes. Branchlets marked with scars of fallen leaves. **Leaves** alternate, bright green, shining, elliptic, 5-10 cm x 3.7-5 cm, pointed, with a toothed margin, on 2-2.5 cm long stalks. **Flowers** minute, yellowish-green, fragrant; male in crowded catkin-like spikes 3.7-5 cm long, female spikes fewer, more slender and shorter than the male. Male flower stalkless, in the axils of rounded bracts. Sepals 3, stamens with long filaments. Female flowers stalked. **Fruit** a 3-lobed capsule 0.7-1.2 cm or more in diameter, smooth. **Seeds** globose, smooth, 0.5 cm in diameter. **Distribution:** Tidal forests and mangroves in the Sunderbans, Andaman and Nicobar Islands, Bangladesh, Myanmar and Sri Lanka. **Phenology:** Leafless April-May, leaves turn bright yellow or red before falling. **Flowers** July. **Fruit** ripens August-September and germinates immediately after falling. **Miscellaneous:** Wood is used for making toys, furniture, etc. On the west coast fishing nets are supported by floats made from its roots. A fuel-wood with high calorific value. It is called the blinding tree on account of the irritating milky juice in the stem, which is said by wood-cutters to cause blindness if it enters the eyes. The generic name *Excoecaria* refers to this poisonous sap.

99. BISHOP WOOD, TIGER TREE
Bischofia javanica Blume

(Family: Euphorbiaceae)

Plate 10

Hindi *Kaen*; Assamese *Uriana*; Tamil *Thondi*.

Field Identification: Bark dark brown or grey, peeling off in angular scales, cut reddish with red juice. Leaves alternate, leaflets 3-5, 7.5-15 cm, margin round

Fig. 55

P. SHARMA

S.No. 98. *Excoecaria agallocha*. 1) Leaves and inflorescence x 1. 2) Flowering twig x 3. 3) A flower x 5. (From *Naithani* 1223).

toothed. Flowers small, greenish yellow, in terminal panicles. Unisexual, male and female on different trees. Fruit globose, 1.2 cm across. Seeds 3 or 4, shining, 3-angled. **Description:** A large quick-growing tree, evergreen in moist regions, deciduous in comparatively dry regions. **Bark** dark brown or grey, peeling off in regular scales, blaze soft and pink; red juice oozes out on cutting. **Leaves** alternate, shining, dark green, leaflets 3-5, common leaf stalk 5-15 cm, leaflets 7.5-15 x 3.7-7.5 cm, pointed, margin round toothed, stalk of leaflets 1.2 cm long. **Flowers** small, greenish yellow, on a branched terminal panicle, unisexual, male and female on separate trees, without petals. Male flowers, sepals falling off. Ovary, 3-4 celled. **Fruit** globose, fleshy, 1.2 cm across, brown, cover parchment-like. **Seeds** 3 or 4, smooth, shining, 3-angled, 0.4 cm long. **Distribution:** Sub-Himalayan tracts Yamuna eastwards to Eastern Himalaya, Assam, Andamans, Bangladesh to Java, Western Ghats, common in the Nilgiris. A widely distributed tree characteristic of river banks and swamps. **Phenology:** Evergreen in moist regions, in comparatively drier regions, nearly leafless briefly during winter. Leaves turn red before falling. **Flowers** March-April. **Fruit** ripens December-February. **Miscellaneous:** Owing to the soft juicy cortex these trees are favoured by tigers for cleaning their claws and in places where tigers abound the bark of trees is often deeply scored with the claw-marks. The popular name tiger tree, now suggested, is therefore very appropriate. A beautiful wood that does not warp. Wood red with strong odour of vinegar. It is durable under water and is used for railway sleepers in Assam and also for making bridges.

100. PUTRANJIVA
Putranjiva roxburghii Wall.

(Family: Euphorbiaceae)

Hindi, Bengali, Gujarati *Putranjiva*; Marathi *Putujan*; Tamil *Irukolli*; Telugu *Kudrajiva*; Malayalam *Pongalam*.

Field Identification: Small to medium sized evergreen, branches drooping. Leaves glossy dark green, 5-9 cm. Fruit *c.* 1.5 cm, egg-shaped, light greenish-white, hairy, on 1.2 cm long stalk, stone hard. **Description:** A graceful medium-sized evergreen tree with drooping branches. **Bark** grey, smooth, marked with horizontal lines of white specks, inner bark yellow. **Leaves** alternate, leathery, dark green, glossy, elliptic, pointed with a wavy margin, faintly and distantly toothed, 5-9 x 2.5-3.2 cm, leaf stalk 0.6 cm long. Male and female flowers borne on the same tree or on different trees. Male flowers small, yellow, in dense axillary clusters, calyx split into 3-5 segments, stamens 3. Female **flowers** green, single or a few together, calyx 5-6 cleft, falling off after flowering. Ovary hairy, 3-celled. Fruit 1.2-1.5 cm long, elliptic, pointed at tip, greenish-white, hairy, on 1.2 cm long stalk, stone very hard, wrinkled, pointed, **seed** one, occasionally two. **Distribution:** Sub-Himalaya from Ravi eastwards through Nepal to Myanmar, up to 700 m and in Sri Lanka. In evergreen forests and on the banks of streams. Stunted in drier regions. **Phenology:** New leaves in April. **Flowers** March-May. **Fruit** takes one year to ripen in winter about March. **Miscellaneous:** Some authors include Putranjiva under Dryptes. Fruit eaten by deer. It is used as an ornamental tree for parks and avenues; it makes a fine hedge after pruning. The nuts are strung in rosaries in India and Sri Lanka. Wood used in tool handles. **Etymology:** Botanists have adopted the Sanskrit name as a botanical name for this tree. It means 'Life of the child', the bony nuts are strung into rosaries which are worn by children in the belief that they ward off evil, *roxburghii* is after an early British botanist in India, William Roxburgh.

101. WEST HIMALAYAN ELM
Ulmus wallichiana Planchon

(Family: Ulmaceae)

W. Him. *Kain, Maral, Emroi*

Fig. 56

Field Identification: Leaves alternate, elliptic-obovate, long, pointed, double toothed, rough, *c.* 8-10 cm long, lateral veins parallel, 15-20 pairs. Flowers numerous, 3 mm long, in globose heads. Fruits winged, elliptic, 1.5 cm across, with a central seed. **Description:** A large deciduous tree with rough grey deeply furrowed bark, peeling off in diamond-shaped scales. **Leaves** elliptic to obovate, long, pointed, double toothed, rough above when mature, usually 8-10 cm long, lateral nerves 15-20 pairs, parallel, each ending in a large tooth. Leaf stalk 0.6 cm. **Flowers** numerous, *c.* 3 mm long, densely fascicled, forming globose heads in the axils of the fallen leaves, perianth 5-6-lobed. **Fruit** a nutlet surrounded by a papery wing 1.5 cm across. **Distribution:** Western Himalaya and Nepal at 1800-3000 m. In broad-leaf forests and moist ravines. **Phenology:** Leafless in winter. New leaves April-May. **Flowers** immediately before leafing. **Fruit** ripens May-June. Much of the seed is barren. **Miscellaneous:** Branches lopped frequently for fodder, timber yellowish-brown in colour with a handsome grain and takes excellent polish. Bark contains a strong fibre used for ropes.

Fig. 56

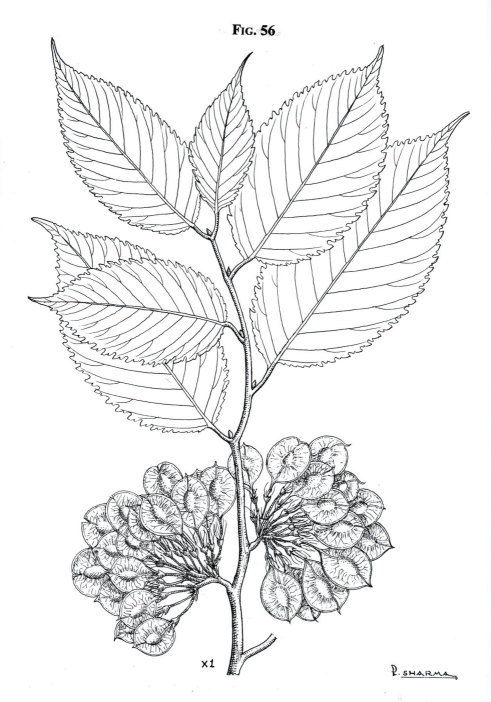

S.No. 101. ***Ulmus wallichiana.*** 1) Double toothed leaves, fruits with circular papery wings x 1. (From *Khan* s.n. and *Hole* s.n.).

Fig. 57

S.No. 102. ***Ficus benghalensis***. 1) A twig showing leaves and young figs x $\frac{1}{2}$.
2) A ripe fig. 3) Cross section of fig x 1.

102. BANYAN
Ficus benghalensis L.

(Family: Moraceae)

Hindi, Bengali *Bar*; Gujarati, Marathi *Vad*; Telugu *Marri*; Tamil *Al*; Kannada *Ala*; Malayalam *Aal maram*; Burmese *Pyin-vaung*.

Fig. 57

Field Identification: A huge spreading tree, with aerial roots descending from the branches which enter the ground, and thicken and become trunk-like. Leaves leathery, oval-shaped, 8-20 cm with thick leaf stalks 2.5-5 cm long. Figs in pairs, globose, 1.2-2 cm, red when ripe, stalkless. **Description:** A large evergreen tree, branches spreading, sending down to the ground numerous aerial roots which afterwards thicken and become trunks and support the crown. **Leaves** thickly leathery, oval-shaped, round at the tip, base heart-shaped, or rounded, blade 8-20 cm, leaf stalk thick, 2.5-5 cm. **Figs** in pairs, globular, stalkless, 1.2-2 cm, red when ripe. Male **flowers** near the mouth of the figs, with one stamen. Sepals 4, lance-shaped. Gall flower: sepals like male, style short. Female flower: perianth short and style longer than in the gall flowers. The banyan and almost all figs have a milky juice and 2 large scales (stipules) which cover the leaf bud. They fall off, leaving a scar. **Distribution:** Sub-Himalayan forests up to 1200 m and the western peninsula. Trees with very large spreading crown are found in southern Maharashtra and in North Kanara. **Phenology:** Evergreen. Leafless in dry localities briefly in the hot season. **Fruits** April-June. Also in October-November. **Miscellaneous:** The leaves are fodder for elephants and camels. In some localities the felling of banyan is forbidden within a mile of the camping ground, in order to ensure supplies of elephant fodder. The figs provide food for a variety of animal life, including man in times of adversity. The leaves are made into plates. The banyan is sacred to the Hindus. The great banyan of Calcutta has 1000 trunks. The canopy covers 4 acres, and a walk round the tree is about 400 m. The specimen in the Botanical Garden, Calcutta, is one of the largest in the world. It was ascertained to have grown from seeds dropped by birds in the crown of a date palm in 1782, therefore it is over 200 years old. Banyan and other figs begin life as epiphytes developing from seeds dropped by birds. They may strangle the host tree, which sometimes dies, but the two can co-exist for a long time. **Etymology:** *Ficus* is the Latin name for the fig tree, *benghalensis* refers to Bengal.

154

103. PIPAL, BODHI or BO TREE
Ficus religiosa L.

(Family: Moraceae)

Hindi, Punjabi, Gujarati *Pipal*; Marathi *Pimpal*; Malayalam *Arasu*; Bengali *Ashuvatham*; Tamil *Arasamaram*; Burmese *Nyaung bawdi*.

Fig. 58

Field Identification: Large tree without aerial roots. Leaves heart-shaped 11-17 x 7.5-12 cm, shiny on the upper surface, apex drawn into a tail-like tip. **Description:** A large tree with a fluted trunk and spreading branches without aerial roots. Epiphytic when young. **Bark** smooth, pale grey. **Leaves** alternate, 11-17 x 7.5-12.5 cm, very shiny on the upper surface with a heart shaped base, abruptly narrowed at the apex into a long tail-like tip, leaf-stalk slender 7.5-10 cm long, which causes leaves to sway in a breeze. **Fig** globular 1.3 cm, in stalkless pairs, dark purple when ripe. Male **Flowers** stalkless, scanty at the mouth of the fig, sepals 3, stamen one, stalkless or stalked. Gall and fertile flower, stalkless or stalked, sepals 5. As the fig of 'pipal' is small the insect associated with it is smaller than insects associated with species with bigger fruits. Each species of *Ficus* usually has its particular insect. The insect cannot live without the fig, and the *Ficus* tree cannot produce seed without the insect. The insects enter through a hole at the tip of the fig and lay eggs in the flowers, which are known as gall flowers. The eggs hatch and mature into new insects which on leaving their home become dusted with pollen from the male flowers. They then make their way into another fig and repeat the process, thus ensuring pollination of other female flowers. Figs of pipal turn dark purple on ripening and are much sought after by birds, notably by flocks of starlings in April, also by green pigeons and monkeys which aid in the dispersal of seeds. **Distribution:** Himalayan foothills from Pakistan through India to Bhutan to Myanmar. It has run wild throughout the Subcontinent. **Phenology:** In North India new leaves appear about March. Young leaves are attractive with a reddish tinge. Old leaves are shed in the winter. **Fruit** ripens in April, also in October-November. **Miscellaneous:** Pipal, like banyan, begins life as an epiphyte developing from seeds dropped by birds in forks of other trees, sometimes in the trunk of a date palm. The two co-exist for a long time, perhaps one hundred years or more. Pipal sometimes strangles the host tree, which gradually dies; it is destructive to buildings. It is a long lived tree. A tree in Sri Lanka is reported to be over 2200 years old (Champion, H.G. & A.L. Griffith, *Manual of General Silviculture for India*, OUP, 1948, pp. 99). **Mythology:** This is the Bodhi Tree under which Buddha sat and received enlightenment at Bodh Gaya. A sapling of the original tree from here was taken to Sri Lanka by Mahindra, son of emperor Ashoka and flourishes

FIG. 58

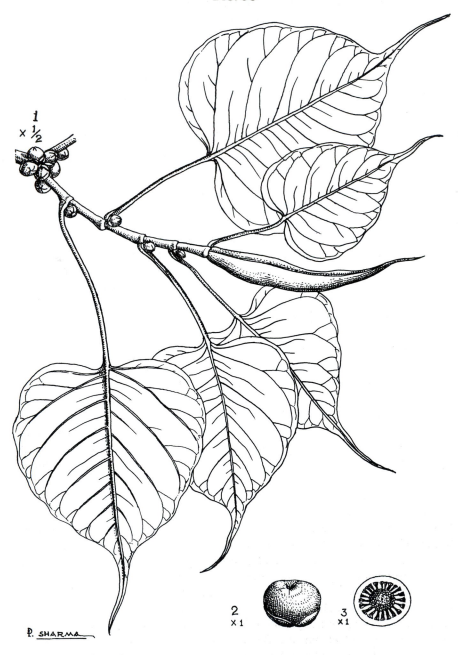

S. No. 103. *Ficus religiosa.* 1) Twig with leaves and young figs x ½. 2) Ripe fig x 1.
3) Cross section of fig x 1.

FIG. 59

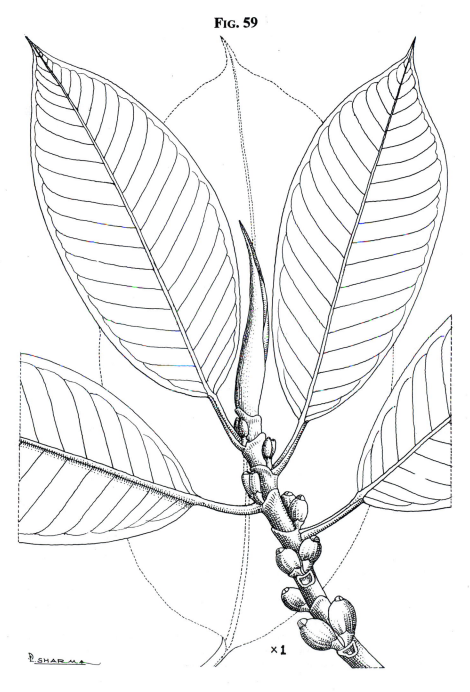

×1

S.No. 104. *Ficus elastica*. 1) A twig with foliage, a long apical stipule and figs x 1.

there today. It is one of the most sacred trees of India, Nepal and Sri Lanka and is venerated both by the Hindus and Buddhists and is often planted near temples and monasteries. Hindus have a strong religious objection to felling it. It does much harm to buildings, sending its roots into crevices. It is a most useful shade tree with a dense crown and is planted at places where the main requirement is to provide shade for weary travellers. **Etymology:** *Ficus* is the Latin name of the fig, *religiosa* in Latin means pertaining to religion and is given as the tree is sacred.

104. ASSAM RUBBER TREE, INDIA RUBBER TREE
Ficus elastica Roxb.
(Family: Moraceae)

Bor. Assamese *Attah*

Fig. 59

Field Identification: Giant buttressed tree with an immense crown with aerial roots. Leaves dark green, leathery, glossy, 12.5-25 cm abruptly pointed at the tip, lateral nerves many, parallel. Stipules pink, up to 15 cm. Figs oblong, 1.2 cm, greenish-yellow when ripe. **Description:** A gigantic, evergreen, much buttressed or fluted tree with a dense crown, sending down numerous aerial roots from the branches. **Bark** greyish or reddish brown, fairly smooth but scurfy. **Leaves** dark green, leathery, shiny, elliptic, abruptly and shortly pointed, blade 12.5-25 cm; midrib prominent, with numerous parallel lateral nerves, leaf stalk 1.2-6.2 cm; **Stipules** pink, up to 15 cm long. Figs enclosed when young by hooded bracts which fall off leaving a saucer-shaped base under the fruit. **Fruit** (Fig) stalkless, ovoid-oblong 1.2 cm long, greenish yellow when ripe. **Seeds** numerous, small. **Distribution:** East Himalayan foothills from Nepal eastwards through Sikkim-Bhutan, Arunachal Pradesh, Assam and Khasi Hills up to 1200 m. Conspicuous in the evergreen forest towering over the surrounding trees. **Phenology: Fruit** ripens by May, falling throughout the rains. **Miscellaneous:** The plant starts life as an epiphyte, germinating on the branches of tall forest trees, grows slowly at first, but soon sends down aerial roots, and when these have taken root, begins to make rapid progress. Eventually the tree stands on its own and attains a height of 29-58 m. It has been in cultivation in Assam since 1874 and held out promise of a great future. It was soon superceded by para rubber *Hevea brasiliensis* which, age for age, produces higher yield per hectare. The figs are eagerly sought after by birds and the bulk of them are devoured on the tree. The trees suffer damage from elephants and deer. The tapping of Assam rubber ceased after introduction of the superior para rubber. It is now used as a shade tree or as a dwarf indoor pot plant because of its handsome foliage. Propagation from seeds sown from June to September or by cuttings and layerings.

158

105. CHAPLASH
Artocarpus chaplasha Roxb.

(Family: Moraceae)

Bengali *Chaplash*; Nepali *Lutta*; Burmese *Thaungpeinne*.

Fig. 60

Field Identification: Large tree. Leaves broadly ovate, 15-25 cm, rough beneath, with short stiff hairs. Leaves of young trees up to a metre long with lobed and serrated margin. Male and female flowers in globose heads. Fruit globose, fleshy, tubercled 7.5-10 cm across. **Description:** A large deciduous tree with a spreading crown. **Bark** brown, rough with tubercles and small fissures, peeling off in large round flakes and exuding a milky juice when cut. Young shoots clothed with stiff hairs. **Leaves** broadly ovate, thinly leathery, 15-25 cm long, leaf stalk 0.6-1.2 cm, undersurface of leaves rough with minute stiff hairs. Leaves of young trees are totally different from leaves of mature trees. In the former they are up to a metre long with lobed and serrated margin. Stipules large, clasping the stem. **Flower** heads male and female separate, globose, on 3.7-5 cm long stalks. **Fruit** globose, fleshy, tubercled 7.5-10 cm across. **Seeds** few 1.2-1.4 cm long. **Distribution:** Sub-Himalayan tracts from Nepal eastward, Khasi Hills, Andamans, Nicobars, Bangladesh and Myanmar. **Phenology:** Leafless briefly in the hot season. **Flowers,** the globose flower heads appear in March-April. **Fruit** ripens June-August. **Seeds** lose viability quickly. **Miscellaneous:** Fruit eaten by birds and monkeys and seeds get dispersed through them. Elephants greatly relish the leaves of chaplash. The yellowish wood is durable, used in canoes, boats and furniture. **Etymology:** The name comes from the Greek, *artos*, bread, and *karpos*, fruit, the fruits are edible.

106. JACK TREE
Artocarpus heterophyllus Lam.

(Family: Moraceae)

Syn. *A. integrifolius* auct non L.f

Hindi *Kathal*; Marathi *Phanas*; Bengali *Kanthal*; Tamil *Pilapalam*; Malayalam *Plavu*; Burmese *Peinne*.

Fig. 61

Field Identification: Tree with a short trunk and a dense crown. Leaves dark green, leathery, 10-20 cm, elliptic-oblong. Male flowers on cylindrical receptacles

FIG. 60

S.No. 105. *Artocarpus chaplasha*. 1) A twig with leaves and a male flower head x 1.
2) Fruit x $\frac{3}{4}$.

FIG. 61

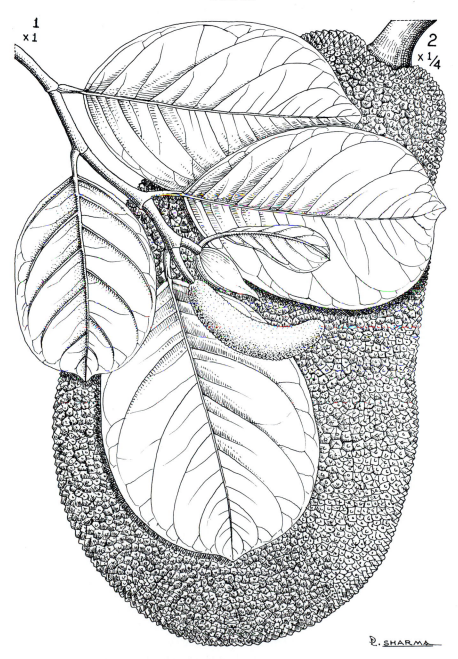

S.No. 106. *Artocarpus heterophyllus*. 1) A twig showing a male receptacle and
leaves. 2) Fruit (much reduced).

up to 15 cm long, female flowers on ovoid-oblong receptacles. Fruit very large, 30-150 x 15-30 cm, yellowish green, with conical tubercles. Seeds ovoid-kidney shaped 2.5-3.7 cm, surrounded by yellow pulp. **Description:** A large evergreen tree with a short thick trunk and a dense round crown. **Bark** dark brown and rough, inside pale pink, on blazing milky juice oozes out. **Leaves** dark green, thick, leathery, shining, elliptic or oblong 10-20 cm, leaf stalk 1.2- 2.5 cm. Stipules large falling off early. Flower heads in bud enveloped in large stipular deciduous sheaths. **Flowers** numerous, yellowish-green, male on stout cylindrical 5-15 cm long receptacles, female flowers on ovoid oblong receptacles. **Fruit** large yellowish green, 30-150 cm x 15-30 cm hanging on short stalks from the trunk, the rind of the fruit is covered with conical protuberances. **Seeds** smooth, ovoid, kidney-shaped, 2.5- 3.7 cm surrounded by yellow pulp. **Distribution:** Western Ghats up to 1100 m, cultivated elsewhere in India, Myanmar, Pakistan, Bangladesh, and Sri Lanka. **Phenology: Flowers** December-March. **Fruit** ripens during the rains in July-August. **Miscellaneous:** Fruit ripens during the rains in July-August, is aromatic sweet and fleshy and is eaten raw in S. India and Sri Lanka. The pulp and the seeds are also cooked . The fruit is relished by elephants.

107. HIMALAYAN MULBERRY
Morus serrata Roxb.

(Family: Moraceae)

W. Himalayan *Tut*; Hindi *Kimu, Himu*; Khasi *Dien-soh-tungkhar*.

Field Identification: Large tree, bark reddish or greyish brown vertically fissured. Leaves 5-20 cm, broadly ovate-heart shaped, coarsely toothed, often 3-lobed, hairy beneath. **Description:** A large deciduous tree with reddish or greyish-brown **bark** with vertical fissures, and with broadly ovate heart-shaped, coarsely-toothed, often deeply 3-lobed leaves, thickly hairy beneath. **Leaves** 5-20 cm, with a long-pointed apex, woolly-hairy beneath when young. Male and female **flowers** borne on the same or on separate trees in catkins. Male flowers in short stalked 1-2 cm long spikes. Female purple 1.6-2.5 cm long, spikes, styles very hairy. **Fruit** purple 1.6-2.5 cm long, borne on a stout hairy stalk. **Distribution:** Himalaya from Pakistan through India to C. Nepal at 1,200-2,700 m. **Phenology:** Leafless in winter. New leaves sprout during March-May. **Flowers** March-May. **Fruit** ripens June-July and is greedily devoured by birds, which disperse the seeds. **Miscellaneous:** Heartwood large, yellow or brown, excellent for furniture and carvings. A sacred tree dating back to the 8th century A.D. is seen in Garhwal at Joshimath at *c.* 1900 m. It is estimated to be more than 1200 years old. The massive trunk consists of 3 main buttressed portions 6.5 m in girth. One of the portions to the extreme left is hollow, its circumference is 12 m. It is a male tree (staminate). The locals say that it has never borne fruit. On account of religious sentiments, the tree is well protected and lopping is forbidden. The great religious preceptor Adi Shankaracharya is said to have meditated under this tree (Rau, M.A. *Indian For.* 93(8): 533-534, 1967).

108. UPAS TREE, SACK TREE
Antiaris toxicaria Leschen

(Family: Moraceae)

Hindi, Andamans *Jangli Lakuch*; Marathi *Karwat*; Tamil, Malayalam *Nettavil*; Kannada *Ajjanpatte*; Burmese *Hmyaseik*.

Field Identification: An enormous buttressed tree. Young shoots velvety. Leaves oblong or ovate-oblong 7.5-15 x 3.7-6 cm, pointed, margin toothed or smooth. Flowers are unisexual, both male and female flowers on the same tree. Male crowded on a thick flat receptacle. Female solitary, enclosed in a pear-shaped circle of bracts. Fruit reddish, fig-like, velvety 2 cm long, one seeded. **Description:** A gigantic evergreen buttressed tree towering to 75 m with buttresses of 10 m or more up the trunk. On blazing milky latex oozes out. Young shoots, leaf stalks and midrib yellow, velvety. **Leaves** oblong or ovate-oblong 7.5-15 x 3.7-6.2 cm, pointed at the tip, margins unbroken or toothed, thinly leathery, rough, shining above, paler and hairy beneath, lateral nerves 8-10 pairs; leaf stalks 0.6 cm long. **Flowers** unisexual, both male and female flowers on the same tree; male flowers axillary, crowded on a thick flat receptacle, sepals 4, spoon-shaped, stamens 5-8. Female flowers solitary, enclosed in a pear shaped circle of bracts. **Fruit** reddish-brown, velvety fig-like, fleshy, 2 x 1.2 cm, fruit stalk 1.2 cm long. One seeded. **Distribution:** Evergreen forests of the Western Ghats from Khandala southwards (once common near Yellapur), Coorg, Andamans, Sri Lanka and on the slopes of Pegu Yoma in Myanmar. In the Thaungyin valley (Myanmar), Brandis in 1899 found Upas trees towering over other trees which had a mean height of 60 m. **Phenology: Flowers** September-October. **Fruit** ripens in the cold season. **Miscellaneous:** The milky latex which oozes out after blazing is poisonous. Jungle dwellers tip arrows and spears with this gum to hunt birds and animals. Inner bark yields a good fibre used for making sacks and hence the popular name Sack Tree. Coats and other garments worn by the 'Ghat' kunbis during the 'Holi' festival are made from this bark. In the Andamans the leaves are used as fodder for elephants. It is a famous tree of legend and fairy-tale. It is reported that all the Dutch soldiers, save one, were killed by poisoned arrows during an encounter with the tribes in the East Indies, that birds were said to fall dead from their perch upon its branches and travellers who went to sleep in its shade never woke up. These stories are sensational flights of fancy. The juice does contain two glycosides, antiarin and strychnine, which affect the heart and arrest its action. It is an interesting, stately and quick growing tree, worthy of introduction in botanical gardens and parks.

109. CHINAR, ORIENTAL PLANE
Platanus orientalis L.

(Family: Platanaceae)

Kashmiri *Buna*; Hindi *Chinar*.

Fig. 62

Field Identification: A majestic tree of great height and girth. Bark scaling in flakes. Leaves 8-24 x 12-30 cm, deeply cut into 3-5 or 5-7 triangular, pointed, toothed lobes. Flowers greenish in long stalked hanging globular clusters. Fruiting heads globular 2-3 cm, of many carpels clothed with long bristly hairs. **Description:** A majestic tree up to 30 m high, 12 m in girth and with a wide rounded crown. **Bark** smooth, thin, light grey or greenish, peeling off in large thin flakes. **Leaves** broadly heart-shaped, 8-24 x 12-30 cm broad, deeply cut into 5-7 triangular pointed, toothed lobes; leaf stalk 2.5-7.5 cm long. **Flowers** greenish in long stalked hanging globular clusters. Male in clusters of 2-3, female clusters larger 1-1.5 cm across. **Fruiting** heads globular 2-3 cm, of many carpels, densely clothed with long fine bristly hairs, the broad apex of the carpels narrowed into a long persistent style. One seeded. Seed downy which helps in dispersal. **Distribution:** Temperate W. Asia, now wild in Turkey and Greece. Introduced in the Kashmir valley about 400 years ago. In cultivation in the W. Himalaya. **Phenology:** Leaves fall in October-November. New leaves in April. **Flowers:** April-May. **Fruit** ripens in June-July. **Miscellaneous:** Wood white fringed with yellow or red, used in making boxes and trays which are lacquered and painted. Chinars are likely to become more widespread, being tolerant of pollution. It is a magnificent tree all the year round, more particularly during autumn, when the leaves turn golden yellow to shades of red. Propagated by cuttings. To propagate, tear off twigs 2-3 years old, *c.* 30 cm long, so that a strip of the older branch or 'heel', to which they are attached, remains at their base. The cuttings should be planted about 10-12 cm deep. Reproduction from seed is common but cuttings are preferable. A fast growing tree. **Etymology:** 'Chi Naar' is a Persian word meaning 'What a Fire', because the leaves turn golden yellow to flaming red in late autumn. At Bijbihara near Srinagar, Kashmir, a tree with a circumference of 16 m was recorded and trees 400 years old are in existence. It was introduced to Kashmir by the Mughals about 400 years ago. The Emperor Akbar introduced a grove at Nasim Bagh (near the present University campus) in Srinagar when he took over Kashmir in 1596. Once a Moghul Royal Tree, it is now the state tree of Kashmir and forms a dominant feature of the landscape of the valley.

FIG. 62

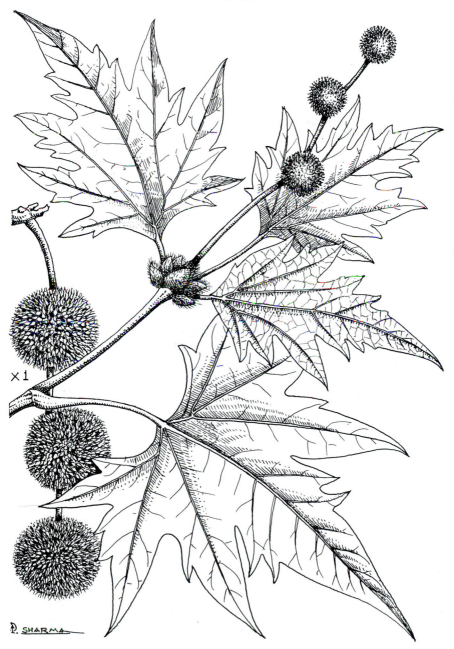

S.No. 109. ***Platanus orientalis***. 1) Young leaves and flowers in globose clusters, later turning into fruits in larger globular clusters.

110. HIMALAYAN WALNUT
Juglans regia L. var. *kamaonia* DC.

(Family: Juglandaceae)

Hindi *Akhrot*, W. Himalaya *Akrut, Akhor, Krot*; Burmese *Thitcha*.

Field Identification: Deciduous tree. Bark dark grey, vertically fissured. Leaves alternate, leathery, leaflets 5-13, elliptic ovate, opposite or sub-opposite, 7-20 cm long, aromatic when crushed. Male catkins hanging, 6-12 cm long, stamens 10-20; female flowers small. Fruit a 5 cm long green drupe with a hard shell with 2 wrinkled edible nuts. **Description:** A large deciduous tree up to 25 m in height and 1.5 m in diameter. Cultivated trees are smaller. **Bark** dark grey, vertically fissured. **Leaves** compound, leathery with 5-13, elliptic-ovate, opposite or sub-opposite leaflets, pointed, 7-20 cm long, aromatic when crushed. **Flowers** Male catkins hanging singly or in pairs 6-12 cm long; stamens 10-20 or more; female flowers very small, 1-3, opposite the terminal leaf. **Fruit** a large 5 cm long green drupe with a hard shell containing 2 wrinkled edible nuts or seeds. **Distribution:** Along the Himalaya from Pakistan through India to Bhutan and hills of Upper Myanmar, mostly at altitudes of 1170-2300 m. **Phenology:** Leaf fall September-November. New **leaves** March-April. **Flowers** March-April. **Fruit** ripens in September. **Miscellaneous:** The nuts are relished by monkeys, squirrels and birds, particularly nutcrackers. Bears also collect the nuts. The winter quarters of a black bear was found to be stocked with walnuts. The nuts are a popular dry fruit. The variety 'Kagazi' (paper shelled) of Kashmir, with a thin shell, is easily breakable between the finger and thumb of one hand. The wood is used for making fine furniture and rifle butts. The burrs which develop on the trunk are valuable for veneers. **Etymology:** *Regia*, Latin "royal" was applied because of the delicious flavour of the nuts.

111. SILAPOMA
Engelhardia spicata Leschen ex Bl.

(Family: Juglandaceae)

Hindi *Silapoma*; Bengali *Bolas*; Nepali *Mauwa*; Assamese *Wakgru*; Lepcha *Sugreot-kung*.

Field Identification: Small to large tree. Leaflets 6-10 paired, rachis 10-30 cm, leaflets oblong to oblong-lance-shaped, minutely toothed, 8-15 x 5-6 cm. Male and female flowers on the same tree. Male spike 3.5-8 cm. Female flowers on long hanging spikes, bracts produced into a wing. Fruit a small round nut with a 3-lobed 5 cm long papery wing. **Description:** Trees 12-30 m, trunk often buttressed in the Eastern Himalaya. **Bark** smooth grey. Leaflets 6-10, paired, leaf rachis

166

10-30 cm, leaflets oblong to oblong-lance shaped, minutely toothed, 8-15 x 5-6 cm, stalks of leaflets up to 5 mm. Male and female flowers on the same tree. Male spike solitary, 3.5 to 8 cm, borne on leafless branches. Female flowers on long hanging spikes, bracts produced into a wing. **Fruiting** clusters long dense, with numerous overlapping fruits; fruit a small globular nut with a 3-lobed papery wing to 5 cm.

— Leaflets glabrous, subsessile; fruiting spikes 20-45 cm long; rachis without hair. var. *spicata* .... 1.

— leaflets hairy beneath, sessile or subsessile; fruiting spikes 10-30 cm long, rachis hairy or without hair.... var. *colebrookeana* .... 2.

Distribution: Himachal Pradesh to Arunachal Pradesh, Manipur and Myanmar up to 1,400 m. **Phenology:** Leafless for a short time during December-February. New **leaves** and **flowers** March-April. **Fruits** May-June. Also flowers at different times from September-May. **Miscellaneous:** The tree produces root suckers which help in stabilising the soil and help in checking erosion on unstable hillsides. Bark is used in tanning.

112. HORSE TAIL TREE, CASSOWARY TREE, WHISTLING PINE
Casuarina equisetifolia J.R. & G. Forst

(Family: Casuarinaceae)

Hindi *Jungli-saru*; Bengali *Jau*; Marathi *Suru*; Gujarati *Vilayati Saru*; Telugu *Saruguda*; Malayalam *Kataadi*; Tamil *Chavukku*.

Fig. 63

Field Identification: Large tree with conifer-like drooping branchlets. Twigs jointed, 6-8 ribbed bearing tiny scale-like leaves set in whorls at the nodes. Male flowers in long spikes, female in compact clusters 1 cm across. Fruit a woody cone *c.* 2 cm across. Seeds winged. **Description:** A large evergreen tree up to 30 m with drooping feathery branchlets looking like a conifer. A close look reveals that the twigs are jointed and 6-8 ribbed, bearing tiny scale-like leaves set in whorls round the nodes. **Bark** peeling off in longitudinal strips. **Flowers** are unisexual, the male growing in long spikes and the female arranged in compact clusters about 1 cm across. **Fruit** a woody cone *c.* 2 cm across containing a number of winged seeds 0.5-0.8 cm long. **Distribution:** Littoral tree on coastal sands in the Andamans, Nicobars and Bangladesh. Planted in the Subcontinent along the coast. **Phenology:** Flowers twice a year February-April and September-October. **Fruit** ripens June-December. **Miscellaneous:** Widely planted on the coastline of the Subcontinent and in Sri Lanka to reclaim sand dunes and check erosion. It is one

FIG. 63

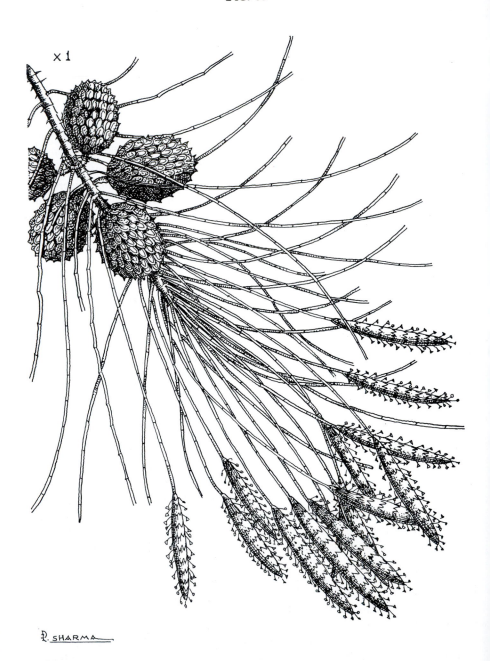

×1

R. SHARMA

S.No. 112. ***Casuarina equisetifolia***. 1) Shoots with flowers. The egg-shaped fruits have twelve rows of seeds (achenes) x 1.

of the best fuelwoods of the world and is the principal fuel wood in the city of Chennai. It is able to fix atmospheric nitrogen through bacteria in its roots. As the tree has a wide distribution from the Andamans to Australia it is very hardy and is capable of growing far inland, as far as Dehra Dun. When clipped it makes a good dense hedge. The timber is used for beams and the bark in tanning. The trees while in fruit produce a whistling sound in a strong   breeze from which the common name is derived. **Etymology:** The name refers to the plumage of the cassowary bird, which resembles the leafless branches of this tree.

113. ALDER
Alnus nepalensis D. Don
(Family: Bctulaccac)

Kunis, W. Him. *Utis*; Lepcha *Kowal*; Burmese *Maibau*.

Fig. 64

Field Identification: Tree with smooth silvery grey bark. Leaves elliptic-ovate 7.5 x 20 cm. Male catkins terminal drooping to 12 cm. Fruiting cones woody, conifer-like, 1.2-2.5 cm long, in erect clusters. **Description:** A medium sized to large deciduous tree with smooth silvery grey **bark** in the open. **Leaves** 7.5-20 cm, elliptic-ovate with a rounded or slightly pointed apex, entire or with a wavy margin, finely hairy beneath when young. **Flowers** Male catkins slender to 12 cm in terminal drooping clusters. Female clusters cone-like, becoming woody. **Fruiting** cones 1.2-2.5 x 0.7 cm in erect clusters. Nutlet with a narrow papery wing. **Distribution:** Himachal Pradesh to Arunachal Pradesh at 1000-3000 m on river sides and wet gullies. Often gregarious. **Phenology: Flowers** September-November. **Fruiting** cones ripen February-March, remaining long on the tree. **Miscellaneous:** It is planted to contain landslides and springs up gregariously on new ground. Wood soft and light, suitable for tea boxes. It is used as hooked sticks in rope bridges.

114. HIMALAYAN SILVER BIRCH
Betula utilis D. Don
(Family: Betulaceae)

Plate 11

Hindi *Bhojpatra*.

Field Identification: Tree with irregular trunk. Bark white to brownish, peeling off in thin rolls. Leaves 5-8 cm ovate, pointed, toothed. Male catkins 7.5-13.5 cm drooping. Female spikes solitary 2.5-3.7 cm. Fruiting catkins 2.5-4 cm. Nutlets 2 mm, winged. **Description:** A moderate-sized deciduous tree with an irregular bole. **Bark** distinctive white to brownish, peeling off in very thin horizontal rolls.

169

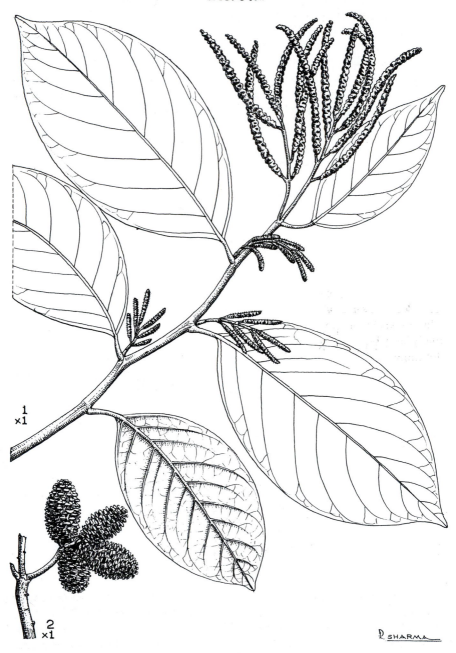

FIG. 64

S.No. 113. *Alnus nepalensis*. 1) A branch with leaves and flowering catkins x 1.
2) Cones in erect clusters x 2. (From *Parker* s.n. and *H.B. Naithani* 262).

Leaves 5-8 cm, ovate fine pointed, irregularly saw toothed, finely hairy when young. **Flowers** Male catkins 7.5-13.5 cm long, drooping, appearing on bare branches, or with young leaves. Female spikes stiff, solitary 2.5-3.7 cm long. **Fruiting** catkins 2.5-4 x 1.2 cm. Nutlets *c.* 2 mm, conspicuously winged, with a 3-lobed much longer bract. **Distribution:** All along the Himalaya from Pakistan through Kashmir to Arunachal Pradesh at 2700-4300 m, forming the uppermost limit of tree growth. **Phenology:** The leaves fall about October, turning a beautiful golden yellow before falling. Leafless during winter. New **leaves** April-May. **Flowers** May. **Fruit** ripens August-October. Winged nuts are dispersed by wind or by drifting or melting snow. **Miscellaneous:** Bark used as paper for writing in ancient times, for water-proofing and roofing houses. Wood used for building in the inner drier Himalaya. Leaves are lopped for fodder.

115. GREY OAK, BAN OAK
Quercus leucotrichophora A. Camus

(Family: Fagaceae)

Syn. *Q. incana* Roxb.

Hindi *Ban, Banj*; Himachal *Ring, Vari*; Garhwal *Phanat*; Kumaon *Ban, Bang.*

Fig. 65

Field Identification: Tree with dull green leaves 6-16 cm long, grey-felted underneath and sharp toothed on the margin. Male flowers in drooping 5-10 cm long spikes. Fruits (acorns) solitary or in pairs 2 cm long. Nut half enclosed in the woody cup. **Description:** A moderate to large evergreen tree with a rounded crown. **Bark** grey, rough with cracks and fissures. **Leaves** leathery dull green, sharply toothed, 6-16 cm long, oblong or ovate-lance-shaped, pointed, with white woolly hairs beneath. Lateral nerves 12-20 pairs parallel. Young foliage lilac or purple tinged. **Flowers:** Male flowers in drooping spikes 5-10 cm long, in clusters of few to several, perianth 4-5 lobed. Female flowers 0.7-1.6 cm long. **Fruit** acorn, solitary or in pairs, 2 cm long. The nut is egg-shaped, half enclosed in the woody cup. **Distribution:** All along the outer Himalaya except the Kashmir valley proper, where the full force of the monsoon is not felt. It is capable of growing on the hottest and driest hillsides and in such situations it is stunted and gnarled, 1200-2400 m. **Phenology:** New **leaves** and **flowers** April-May. **Fruit** ripens August-October 15-17 months after flowering. **Miscellaneous:** Nuts are greedily eaten by bears, monkeys, flying squirrels, birds especially jays and nutcrackers. A sweet exudate, known as Oak Manna,

FIG. 65

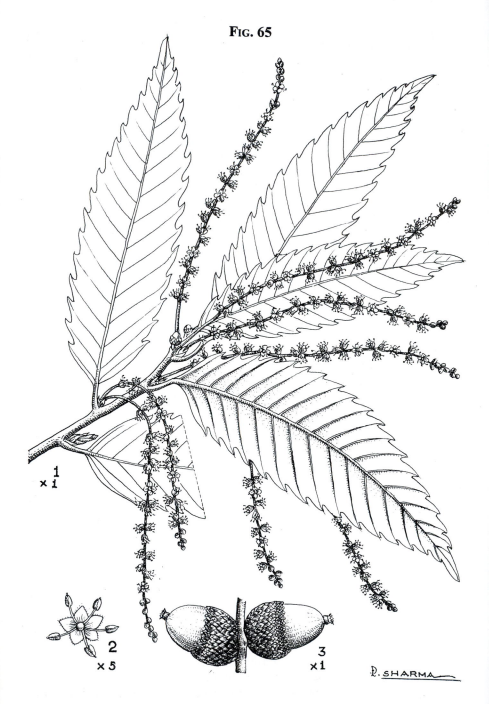

S.No. 115. *Quercus leucotrichophora*. 1) A branch with sharply toothed leaves,
male flowers. 2) Flower x 5. 3) Fruits in pairs, egg-shaped nut in woody cup x 1.

is used in confectionery. Wood is used locally in buildings, gives good fuel. Foliage is lopped for   fodder.

116. BROWN OAK, KHARSU OAK
Quercus semecarpifolia Smith
(Family: Fagaceae)

W. Him. *Kharsu, Banjar, Kreu, Khor*; Bhut. *Khusra.*

Field Identification: A large tree of high altitudes. Leaves 5-10 cm, spinous toothed, stiff and leathery on young trees, entire in old trees, rusty hairy beneath. Male spikes crowded, 5-10 cm, hanging. Acorn solitary 2.5 cm, black when ripe, cup 1.2 cm across enclosing 1/3 of the nut. **Description:** A large evergreen tree. **Bark** dark grey, rough. **Leaves** elliptic-oblong 5-10 cm long, spinous toothed, stiff and leathery on young trees, often with entire margin on old trees, dark green glossy above and generally with rust-coloured hairs beneath, often nearly hairless on old trees. **Flowers** Male spikes crowded, 5-10 cm long, hanging, stamens indefinite. Female spikes short. Fruit (acorn) solitary, nut globose 2.5 cm, black when ripe, cup 1.2 cm in diameter enclosing a third of the nut. The foliage has coppery or brownish tinge in autumn and winter and hence the popular name Brown Oak. At this time Kharsu forests look spectacular under clear blue skies. **Distribution:** All along the Himalaya, absent in the Kashmir valley proper, through Himachal to Arunachal Pradesh and the Manipur border at high altitudes from 2100-3800 m. **Phenology:** New leaves and flowers May-June. **Fruit** ripens in August-September, 15 months after flowering. **Miscellaneous:** Bears are extremely fond of the acorns. Wood is used for buildings, is an excellent fuel and makes very good charcoal. The bark contains much tannin. Foliage is lopped for fodder. Leaves are suitable for feeding caterpillars of the silk-moth *Antheraea pernyi* introduced from China.

117. BUK OAK
Quercus lamellosa Smith

(Family: Fagaceae)

Lepcha *Buk*; Nepali *Bujrat Shalshi*; Burmese *Thite*.

Field Identification: Large umbrella-shaped tree recognized by its large leaves up to 30 cm with white undersurface, sharply toothed margin like a saw, prominent up to 25 pairs of parallel lateral nerves. Fallen leaves and fruit on the ground aid in instant recognition. Acorns ringed by several lamellate cups up to 7 cm across. Nuts top shaped up to 3.7 cm long, brown and smooth. **Description:** A very large umbrella shaped evergreen tree up to 60 m in height. **Bark** grey with a pinkish tinge in old trees. **Leaves** large up to 30 cm long oblong-lance-shaped with a white undersurface. Leaf margin sharply toothed like a saw, lateral veins parallel up to 25 pairs, prominent beneath. Young leaves silvery hairy or buff hairy beneath. **Fruits:** Very large, stalkless, borne on short spikes with concentrically ridged cups encircling the nut. Nut globular or top shaped 1.8-2.9 x 2.2-3.7 cm brown, smooth. Cup up to 7 cm across. **Distribution:** Arunachal Pradesh, Manipur at 1,600-2,800 m, in moist forests; gregarious and common in wet areas. **Phenology:** Flowers April-May. Fruit (acorns) ripens in November-December of the second year. **Miscellaneous:** Acorns greedily eaten by bears, monkeys, squirrels and pigs. It is the most important oak of the Darjeeling Hills. Wood hard and durable, used for buildings. Excellent fuel. Foliage is lopped for fodder.

118. HIMALAYAN POPLAR
Populus ciliata Wallich ex Royle

(Family: Salicaceae)

Hindi *Safeda;* Pahari *Pipal*; Kumaun *Syan*; Nepali, Bengali *Bangikat*.

Fig. 66

Field Identification: Tree with a clean bole and greyish bark, fissured when old. Leaves heart-shaped, pointed 7-16 cm, margins serrated, on long stalks to 12 cm. Male and female catkins on separate trees. Male catkins hanging to 10 cm, female stiffer, hanging to 22 cm. Fruiting catkins to 30 cm. Capsule 8-12 mm. Seeds 2 mm covered in long silky hair to 0.9 cm. **Description:** A lofty tree with greenish grey vertically fissured **bark** on old trunks. **Leaves** heart-shaped, 7-18 cm long, long pointed, with serrated and glandular hairy margins; leaf stalk 5-12 cm long. Buds sticky and resinous. **Flowers** male catkins 5-10 cm long, hanging with yellow

FIG. 66

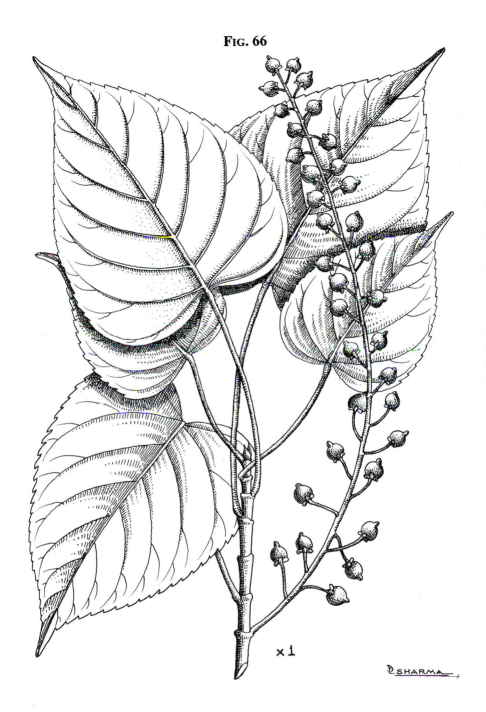

×1

P. SHARMA

S.No. 118. *Populus ciliata*. Leaves, fruits x 1.

bracts tinged with long hairs. Male and female catkins are on separate trees, the former are uncommon. Female catkins are stiff, green and hanging 15-22 cm long. **Fruiting** catkins lax 15-30 cm long. Seeds 2 mm long enveloped in 0.5-0.9 cm long silky hair. **Distribution:** All along the Himalaya from Kashmir to Arunachal Pradesh and beyond to China at 2,100-3,600 m, mostly along streams and in coniferous forests in drier areas. **Phenology:** Starts shedding leaves in September, leafless by November. New **leaves** by end March. **Flowers** early April. **Fruit** ripens in June. The air around Himalayan poplars is full of flocks of down floating about in the breeze. In hill stations it is desirable to plant male trees which are more handsome. **Miscellaneous**: It is fast growing, used for matches and packing cases for apples. Used as fuel in the cold deserts of Ladakh and for water troughs. Leaves provide fodder. Male trees are handsome and are recommended for hill stations as the down of female trees gets into the eyes. **Propagation:** By cuttings in February. **Etymology:** The name *Populus* is derived from *arbor populi,* the peoples' tree, the name the ancient Romans used for this tree; *ciliata* in Latin refers to the hairs on the leaves.

119. INDIAN POPLAR
Populus euphratica Olivier

(Family: Salicaceae)

W. Him., Punjab *Safeda, Bhan*; Ladakh *Hotung, Howdung*.

Fig. 67

Field Identification: Trees from sea-level to 4,000 m. Leaves variable, linear, 7.5-15 cm, shortly stalked on young trees and coppice shoots. Ovate, rhomboid or heart-shaped, 5-7.5 cm long, margin toothed, on long stalks to 5 cm in older trees. Male catkins 2.5-5 cm. Female to 7.5 cm. Capsules lance-shaped 0.6-1.2 cm on slender stalks. Seeds surrounded by silky hairs. **Description:** A medium-sized deciduous tree in the plains, small in the high altitudes of Ladakh, usually gregarious. **Bark** thick, with irregular vertical furrows. **Leaves** variable in shape, those of seedlings, young trees, and coppice shoots linear 7.5-15 cm long, short stalked, those of older trees, circular ovate, rhomboid or heart-shaped 5-7.5 cm, leaf stalks 2.5-5 cm long, leaf margin toothed or lobed, narrow leaves are generally with unbroken margin. **Flowers** in lax catkins. Male catkins 2.5-5 cm long, bracts oblong-lance-shaped, stamens 8-12, on a long slender stalk. Female catkins 5-7.5 cm. **Capsule** lance-shaped 3-valved, 0.6-1.2 cm on a long slender stalk. **Seeds** surrounded by silky hairs. **Distribution:** Along the Indus in Sind and ascending up to 4000 m in Ladakh and W. Tibet, also along river banks in Baluchistan and the Punjab. It has an extraordinary range of latitude, being found from the Altai

176

FIG. 67

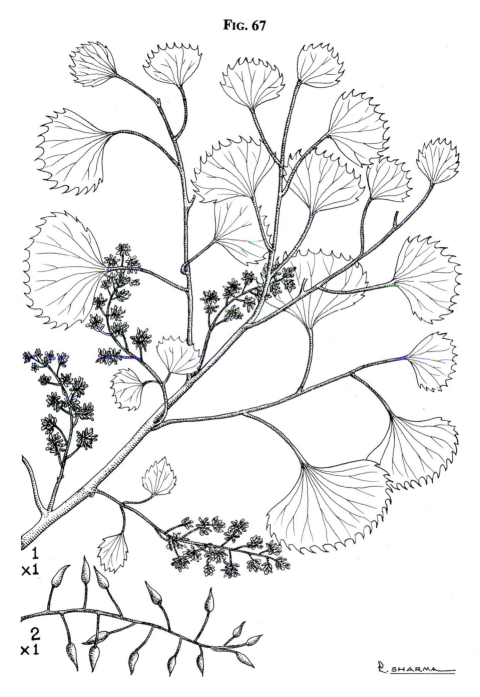

S.No. 119. ***Populus euphratica***. 1) Branch with leaves and male flowers x 1.
2) Fruits x 1. (From *Parker* 62370, *Lambert* s.n.).

Mountains, 45°N to the Equator and also occurs along the river Euphrates. Its outstanding feature is its ability to withstand very high and low temperatures and scanty rainfall. **Phenology:** Leafless from January-March, before falling the **leaves** turn a golden-yellow and look beautiful. **Flowers** January-February. **Fruit** ripens April-June. Seeds with silky hairs are dispersed by wind. **Miscellaneous:** In the cold deserts of Ladakh its bark is browsed by animals and thus prevents wildlife from starving. Wood suitable for matches, turnery and for lacquered articles in Sind, also suitable for shoe heels and cricket bats. **Etymology:** *Euphratica* refers to the river Euphrates as it also occurs along that river, and is known there as the Euphrates poplar.

120. INDIAN WILLOW
Salix tetrasperma Roxb.

(Family: Salicaceae)

Punjabi *Bis*; Hindi *Laila*; Bengali *Panijama*; Telugu *Eeitipale*; Tamil, Malayalam *Vanji*; Assamese *Veh*; Khasi *Jamynrei*.

Fig. 68

Field Identification: Small to medium-sized tree. Leaves lance-shaped minutely toothed, 5-15 cm, on stalks to 2.5 cm. Male catkins 5-10 cm, stamens 5-10. Female catkins 7.5-12.5 cm. Capsules on groups of 3-4, 0.3-0.4 cm. Seeds 4-6. **Description:** A small to medium sized deciduous tree. **Bark** greyish brown rough with vertical furrows. **Leaves** 5-15 cm, lance-shaped, minutely toothed, on leaf stalks to 2.5 cm. Male catkins 5-10 cm, stamens 5-10. Female catkins 7.5-12.5 cm. Capsules in groups of 3-4, 0.3-0.4 cm long. Seeds 4-6. **Distribution:** Throughout the greater part of the Subcontinent along streams. In the Himalaya up to 1,500 m and in the Nilgiris up to 2,000 m. **Phenology:** In North India leafless in the cold season. New leaves and **Flowers** (catkins) February-March. **Fruit** ripens April. In Maharashtra and South India flowers October-November. **Miscellaneous:** The wood is used in gunpowder and twigs in basketry. The Cricket-bat willow *S. alba* ssp. *coerulea*, the best wood for the purpose, was successfully introduced in Kashmir from the United Kingdom. The buyer takes care to select female trees only and reject all male trees. It has a tapering, pyramidal crown with greenish-blue leaves and purple twigs. *S. babylonica* (weeping willow) with graceful hanging branches, is popular in landscaping. Introduced from the Levant, the young foliage has the fragrance of roses.

FIG. 68

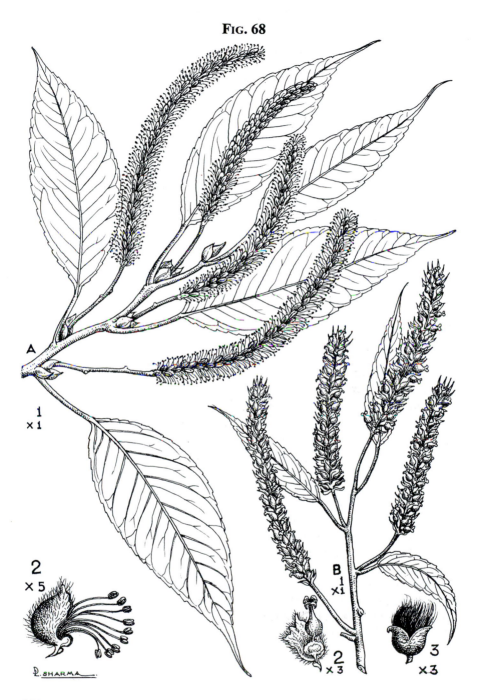

S.No. 120. ***Salix tetrasperma***. **A.** 1) Leaves, male catkins x 1. 2) Male flower x 5.
B. 1) Twig, female flowers x 1. 2) Female flowers x 3. 3) Fruit x 3. (From *Fischer* s.n.).

121. EDIBLE DATE PALM
Phoenix dactylifera L.

(Family: Arecaceae)

Hindi, Bengali, Marathi, Gujarati *Khajur*; Tamil *Perichchankai*;
Telugu *Khajurama*; Kannada *Kharjura*; Oriya *Khorjjuri*.

Field Identification: Palm 30-36 m. Foot surrounded by dense root-suckers.
Foliage feather-like 4.5 m. Male panicle 15-22.5 cm enclosed in a rusty spathe,
flowers sweet scented. Female inflorescence 30-60 cm. Fruit oblong 2.5-7.5 cm,
brown when ripe, pulp edible and sweet. Seed cylindric. **Description:** A tall palm
30-36 m trunk covered with leaf bases of stalks, the foot surrounded by a dense
mass of root-suckers (by which it is distinguished from wild date, *P. sylvestris;*
the latter is also less tall to 15 m). An open crown with feathery greyish green
leaves up to 4.5 m forming a terminal cluster. Petioles spiny. **Leaf** segments are
linear, pointed, 20-40 cm long. Male and female **flower** on different trees. Male
panicles white 15-22.5 cm long enclosed in a rusty spathe, flowers 0.35-0.6 cm, sweet
scented. Female inflorescence 30-60 cm. **Fruit** oblong, 2.5-7.5 cm, yellowish
brown when ripe, pulp fleshy and sweet. Fruit a drupe, whose fleshy and sweet
pericarp is the edible part. **Seed** cylindric furrowed in front. **Distribution:**
Indigenous to the Euphrates and Tigris, and the Oasis of the Sahara, brought to
India during the first Mohammedan conquest in the 8th century. Now wild in
semi-arid Sind, Rajasthan, South Punjab and elsewhere. Introduced in Spain by
the Arabs. **Phenology:** New **leaves** February-March. **Flowers** March-April. **Fruit**
ripens September-October at 6-10 years. **Pollination** is done artificially, effected
by shaking the male inflorescence in the spathe of the female; or the male
inflorescence is removed, spathe slit open and male flowers placed in a basket for
24 hours. The cultivator climbs the female tree on which the spathes have burst
and ties sprigs of male flowers among the female flowers, pollination being thus
effected. **Miscellaneous:** It is a most popular dry fruit. In Arabia the varieties are
known by romantic names as "lady's finger", "pretty maiden's eyes". Leaves are
used for making mats, baskets, fans and leaf stalks as walking sticks, the trunk in
ceilings. Propagation by seed or offset (suckers), the latter are better for propagating
good varieties. The ratio of 3-4 male to 100 female trees is sufficient. The
Khajuraho temples were named after the abundant khajur trees in their vicinity.
Etymology: From the Greek "Phoenix" meaning purple — a name given to the
date on account of its colour which is between yellow and purple, *dactylifera*
refers to the palmate leaves.

122. TALIPOT PALM, FAN PALM
Corypha umbraculifera L.

(Family: Arecaceae)

Marathi *Tali*; Telugu *Sritalama*; Tamil *Kudaippanai*; Kannada *Sritale*; Malayalam *Kudappana, Sitalam, Talippana*.

Field Identification: A tall palm 9-24 m. Leaves very large 2.3-4.6 m across, fan-shaped, cleft into 80 or 100 segments, petiole 5-7 m with spines. Panicle of flowers 6 m tall by 9-12 m wide. Flowers bisexual, calyx 3, petals 3, stamens 6, ovary 3 lobed. Fruit olive coloured, up to 3.7 cm across. Seed globose hard smooth and polished. **Description:** A monocarpic palm i.e. it flowers and fruits once in its life time on attaining maturity and then dies. A large palm 9-24 m high with straight cylindrical trunk 0.5-0.8 m in diameter marked with rings. **Leaves** very large 2.3-4.6 m across, fan-shaped, cleft to about the middle into 80 or 100 linear lance-shaped segments up to 15 cm broad; petiole 5-7 m long armed with dark spines. It spends its whole life in preparation for the supreme act of flowering and fruiting. After many years of growth the talipot palm develops a gigantic terminal bud or spadix over a metre high enclosing the inflorescence. This, in due course, bursts with a loud pop and releases a majestic and gigantic panicle 6 m tall by 9 to 12 m across. The inflorescence containing 60 million flowers is the largest known among flowering plants. After flowering the palm dies. **Flowers** bisexual, small. Calyx 3, broadly lobed, petals 3, stamens six, ovary 3 lobed. **Fruit** a globose drupe, olive coloured 1.8-3.7 cm across with 2 abortive carpels. **Seed** globose very hard smooth and polished. **Distribution:** Malabar, moist monsoon forests of Kumpta and Honavar in North Kanara, gregarious near Gersoppa rain-forests. Middle and South Andaman, Little Coco, Great Coco and Sri Lanka. **Phenology: Flowers** November-January. **Fruit** ripens during the hot season and fall. Flowering takes place between the ages of 17 to 40 years. **Leaves** wither and drop after flowering. Seeds are disseminated by birds, squirrels and porcupines and by rain. **Miscellaneous:** Sacred records of the Singhalese were written on blades of leaves with brass or iron styles. Pith of the trunk is ground into cakes and tastes like white bread. Seeds are used as beads or are coloured and sold as coral. Umbrellas are made of leaves. The broad leaf segments were highly prized for writing upon. Folded leaves are very light and are used by ladies as fans. One leaf shelters 15-20 men against rain and used as tents by soldiers in Sri Lanka.

123. INDIAN DOUM PLANT or BRANCHING PALM
Hyphaene dichotoma (White) Furtado

(Family: Arecaceae)

Field Identification: Palm with repeatedly forking dark grey trunk and fan-shaped leaves. Fruit heart-shaped 8 x 6 cm, brown shining. **Description:** The only palm in India with a branching trunk, sometimes repeatedly, attaining a height up to 15 m. (The allied *H. thebaica* (L.) Mart of North Africa is also a branching palm). **Bark** dark grey. **Leaves** rounded, fan-shaped measuring up to *c.* 91 cm in a terminal crown on each branch, palmate; lobes linear lance-shaped divided 3/4 almost to the base into up to *c.* 40 segments; petiole *c.* 1 m long armed with black spines. **Flowers** Male and female flowers on different trees, sometimes hermaphrodite trees are found. Male spadices *c.* 100 cm long, spathe covered with woolly hairs difficult to remove, stamens six. **Fruit** 8 cm long by 6 cm broad, shining brown, heart shaped. **Seed** 3.5 x 2.5 cm. **Distribution:** Endemic in the island of Diu where it is gregarious on the coastal sands from where it has spread to adjacent Gujarat and elsewhere on the west coast. That it is endemic to Diu is evidenced by the fact that it is not mentioned in *Indian Trees* by Brandis (1906), Talbot's *Forest Fl. of Bombay Pres.* 1911, and Troup's *Silviculture of Indian Trees* 1921. **Phenology:** Flowers October-November. Fruit ripens December onwards. **Miscellaneous:** Leaves as mats and fodder. Trunk as beams, rafters and firewood conservation. It has been listed as Endangered, its use for firewood and fodder has been halted. **Gardening:** From seed, in sandy beds. **Etymology:** From the Greek *hyphaino* = network, referring to the fibres of the fruit, *dichotoma* in Latin means branching.

124. TODDY PALM, PALMYRA PALM, INDIAN FAN PALM
Borassus flabellifer Linn.

(Family: Arecaceae)

Hindi *Tar, Tae*; Marathi *Tad*; Tamil *Panai*; Malayalam *Pana*; Telugu *Tadi.*

Field Identification: Tall columnar dioecious palm with large fanshaped leaves up to 1.4 m in diameter. Drupes heavy, orange coloured 15-20 cm across in clusters. **Description:** A tall, erect, columnar palm up to 30 m x 60 cm in diameter. **Trunk** very dark grey to almost black, smooth, slightly wider above the middle. **Leaves** fan-shaped 0.9-1.4 m in diameter, shining, dark green, divided into 60-80 linear lance-shaped segments, folded along the midrib, petiole 0.6-1.2 m long; indented on the margin. Terminal crown of 30-40 leaves. **Flowers** green, male and female on different trees. Spadix large, sheathed with open spathes, male with stout branches densely clothed with bracts, enclosing spikelets of flowers, sunk in cavities of bracts. Female spadix sparsely branched. Male flowers in small spikelets, calyx and corolla 3 + 3, stamens 6. Female flowers globose, 2.5 cm across larger than male, perianth fleshy, enlarging in fruit, ovary 3-4 celled. **Fruit** a drupe, orange coloured, 15-20 cm in diameter, surrounded by an enlarged fleshy calyx, reddish black when ripe. **Seeds** 3-1, fibrous outside, hard shelled. **Distribution:** Dry areas of Tamil Nadu, Andhra Pradesh and Maharashtra where there are extensive groves. On the banks of the Irrawady in Myanmar, all round Sri Lanka. On low sandy plains exposed to the burning sun and the monsoon. **Phenology: Flowers** March-April. **Fruit** ripens April-May and matures in July-August. **Miscellaneous:** Vast quantities of toddy are drawn from groves of the toddy palm. According to connoisseurs toddy tastes like mild champagne, others associate its taste with cider. Sugar and jaggery has been made from it for 4,000 years. Fruits edible, and fibre and cordage are obtained from many parts of the palm. In the Sudan the fruits are buried in pits, allowed to germinate and then eaten as a delicacy. (Colvin, R.C 1939, Agric. Surv. of Nuba Mts. McCorquodal, Sudan). Fresh sap is used as vinegar. In ancient India the leaves were used as writing material, the parallel veins served as ruled note books. Brandis *Indian Trees* (1906), and Troup, *Silviculture of Indian Trees* (1926) treat this as an African Palm, naturalized in India. The African Fan Palm is a distinct species, *B. aethiopum* Mart. Syn *B. flabellifer* L var *aethiopum* (Mart.) Warburg.

125. FISH TAIL PALM
Caryota urens Linn.

(Family: Arecaceae)

Hindi *Mari*; Marathi *Berli*; Gujarati *Shakarjata*; Tamil *Tippili*; Telugu *Jilugujattu*; Malayalam *Anapana*; Burmese *Minban*.

Field Identification: A palm with gracefully curved bright green shining leaves 5-6 m long, with leaflets shaped like the fins of a fish. Characterized by a large 4 m long drooping spadix in the form of a horse tail. Male and female spadix produced alternately on the same tree. **Description:** A fast growing palm up to *c.* 15 m tall, trunk 0.5 m in diameter. Crown thin, of gracefully ascending, curved 5-6 m long, bright green shining leaves; leaflets lobed like the fin of a fish. Characterized by a large 4 m long drooping spadix in the form of a horse tail. Male and female **flowers** on the same tree, male and female spadix usually alternating. Flowers numerous, in threes, the central and lowermost female, which develop later than the male. Male, sepals 3, heart-shaped, petals leathery, stamens *c.* 40. Female flowers like male, with broader sepals, shorter greenish petals. Ovary triangled, roundish, trilocular. **Fruit** *c*. 1.8 cm across, reddish. **Seeds** one or two. **Distribution:** Sub-Himalayan tracts from Nepal eastwards ascending to 1475 m in Assam, Khasi Hills, Manipur, Western Ghats, Eastern Ghats, Nilgiris, Chittagong, Upper Myanmar, and Sri Lanka. **Phenology: Flowers** most of the year during hot and rainy season. The tree attains full size in 10-15 years. When the fruiting is over the tree gets exhausted and dies. **Miscellaneous:** The petioles and base of the leaf sheaths yield Kitul, a strong fibre used for making brushes, fishing lines and strong wiry ropes for tying elephants. The cabbage or terminal bud is edible and the leaves are relished by elephants. The flowering stalks are tapped for toddy. The timber is strong and durable. **Etymology:** From the Greek "Caryotos" nut-like, on account of the shape of the fruit. The specific name *urens* refers to the burning feeling when the pulp is tasted.

126. BETEL NUT PALM, ARECA PALM
Areca catechu L.

(Family: Arecaceae)

Hindi *Supari*; Gujarati *Sopari*; Kannada *Adika*; Tamil *Pakku*; Telugu *Vakka*; Malayalam *Kamugu*; Burmese *Kun*.

Field Identification: A graceful palm 10-30 m with a slender ringed stem 15-22.5 cm in diameter. Fronds 2 m, feather-like. Flowers unisexual, female flowers surrounded by small white scented male flowers. Fruit an ovoid orange berry 3.5-5 cm long with soft fibrous covering. Seed 2 cm in diameter, reddish yellow. **Description:** A graceful single stemmed palm 10-30 m tall with a slender ringed trunk 15-22.5 cm in diameter. It has a crown of big fronds up to 2 m long, feather-shaped with large segments. **Flowers** unisexual, in an inflorescence with branching spadix. Female flowers grow singly surrounded by the smaller white scented male flowers with 6 stamens. **Fruit** an ovoid orange berry 3.5-5 cm long with a soft fibrous covering. **Seed** 2 cm in diameter, reddish yellow. **Distribution:** Introduced from Malaya from time immemorial, it is found in the Western Ghats, eastern India and Bangladesh. **Phenology: Flowers** January to March. **Fruit** ripens October-January. Isolated trees fruit in about 7 years and continue fruiting for 30-60 years. **Miscellaneous:** The betel-nut is chewed with the leaf of *Piper betel*, the pan. 'Supari' collectors climb to the top of the trees, and hop from one tree to another collecting nuts, as in plantations the tree are closely spaced. The spathes are used as a covering leaf of 'cheroots' in Myanmar. They are also used to write upon and as wrappers for parcels. **Etymology:** Derived from the common name used by the people of Malabar.

127. COCONUT PALM
Cocos nucifera L.

(Family: Arecaceae)

Hindi, Gujarati *Nariyal*; Marathi *Narel*; Kannada *Narikelema*; Malayalam *Thengu*; Tamil *Thenai*; Bengali *Narkel*.

Field Identification: A coastal palm with huge 3-4 m long feather-like leaves. Spadix 1-1.7 m long with an inflorescence with drooping spikes with basal female flowers with some male flowers, upper portion densely covered with male flowers. Fruit round, 25-37 cm with a fibrous mesocarp. The seed lies within a hard woody shell with 3 basal pores. Cavity filled with coconut milk. **Description:** A palm 20-30 m tall, trunk marked in rings each year by the scars of fallen leaves and crowned by an apical cluster of huge 3-4 m long feather-like **leaves** with stout stalks. The **flowers** are united in an inflorescence with a stout spadix 1.1-1.7 m long, divided into drooping spikes, bearing at their bases female flowers with a few male flowers, with the upper portion being densely covered with male flowers with 6 stamens. **Fruit** 3 cornered, 25-37 cm long, pericarp thick fibrous, endocarp bony with 3 basal pores. Cavity of endosperm filled with coconut milk before ripening. **Distribution:** It is believed to be of Indo-Malayan origin. Plentiful in the Coco and Nicobar Islands where it is believed to be indigenous. Widespread and cultivated for centuries throughout the tropics in coastal areas. **Phenology:** New **leaves** arise from the centre of the apical cluster from a bud known as "cabbage". They take 1.5 years from first appearance to their full development. **Flowers** dry season. A coconut tree is estimated to produce 400 million pollen grains every year. **Fruit** ripens after 9-10 months. Fruiting commences from 5 to 7 years after planting. **Propagation:** By seed; it reproduces easily because the fruit float from island to island. **Miscellaneous:** The sweet coconut milk is a refreshing drink. Fibrous husk of fruit is used for ropes and mats. The cut flower stalks are tapped for toddy. The kernel is edible. Oil from the kernel is used commercially. Leaves are used as thatching material and for making mats. **Etymology:** The name is derived from the Portuguese Coco or monkey, alluding to the nut being like a monkey's head.

128. SCREW PINE, KEORA
Pandanus odoratissimus Linn.

(Family: Pandanaceae)

Hindi *Keora*; Bengali *Keya*; Gujarati, Marathi *Kevada*;
Malayalam, Telugu *Mugali*; Tamil *Tilai*; Burmese *Satthapu*.

Fig. 69

Field Identification: Palm-like trees or shrubs on stilt roots. Leaves sword-shaped to 1.5 m long with spines on the margin and on the midrib beneath. Fruit oblong, 15-25 cm in diameter, drooping, pineapple-like, orange or scarlet with 50-60 top-shaped drupes up to 15 cm long. **Description:** Evergreen palm-like tree or shrub up to 6 m high, much branched, resting on strong aerial roots. **Leaves** glaucous green, sword-shaped 0.9-1.5 m with a long pointed tip armed with spines on the edges pointing forward and on the midrib pointing forward and backward. **Flowers** Male flowers in the form of loose conical spadix 25-50 cm long with numerous sub-sessile cylindric spikes 5-10 cm long, enclosed in white, very fragrant spathes. Staminal column 0.6-1.2 cm long, anthers cuspidate. The spadix of the female flowers is solitary, 5 cm in diameter. **Fruit** oblong or globose, 15-25 cm in diameter, drooping, yellow or red, formed of 50-60 drupes, pineapple like in appearance. Each drupe consisting of 5-12 carpels; stigma short, kidney-shaped, yellow. **Distribution:** Sunderbans, sea coast of Peninsula on both sides, Andamans, Nicobars, Gt. Nicobar, Coco Islands, Sri Lanka, Myanmar and Bangladesh. Usually forming a belt above high tide mark, very common and gregarious. **Flowering and fruiting:** Hot and rainy season. **Miscellaneous:** The fibre from the leaves is used for nets, brushes, etc. and the pulp of the fruit is edible. Cultivated for its fragrant flowers. The long ivory coloured bracts of the male inflorescence are used in hair decoration. The cultivated variety is used for extraction of 'kewda attar' and 'kewda water' and is highly prized in Indian perfumes.

129. GIANT THORNY BAMBOO
Bambusa bambos (Linn.) Voss

Syn. *Bambusa arundinacea* Willd.; *B. spinosa* Roxb.

(Family: Bambuseae)

Hindi *Kanta Bans*; Marathi, Gujarati *Bans*; Bengali *Ketua*;
Telugu *Bongu-Vedura*; Kannada *Biduru*; Tamil, Malayalam
Moongil; Assamese *Kotoha*.

Field Identification: A tall thorny bamboo. Culm sheaths 30-37 x 22-30 cm are characteristic with black felt of hairs on the blade, and narrow pointed leaves. **Description:** A tall graceful, thorny bamboo. **Culms** bright green, crowded,

FIG. 69

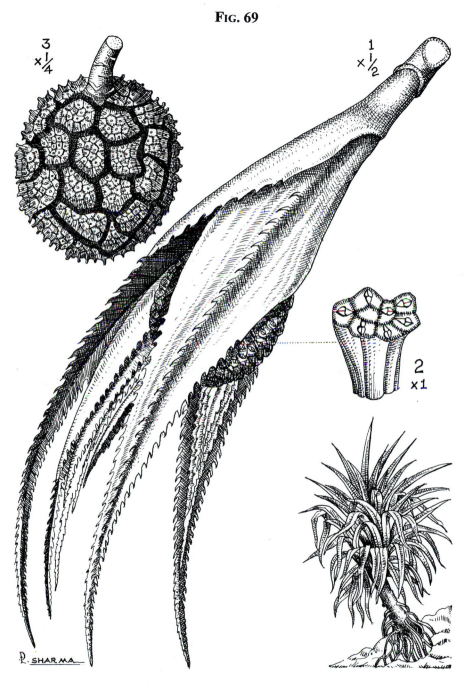

S.No. 128. **Pandanus odoratissimus**. 1) Solitary, terminal female flower x $\frac{1}{2}$.
2) Drupes x 1. 3) Fruit x $\frac{1}{4}$. Screw pine on stilt roots (From *Roxburgh*).

shining, large specimens attaining 28-34 m by 15-17.5 cm in diameter, thick walled 2.5 cm thick, curving at the top; cavity small. **Nodes** with 2 to 3 recurved spines. **Internodes** up to 45 cm long. **Culm sheaths** leathery, orange-yellow when young, 30-37.5 cm by 22.5-30 cm broad, thickly covered with golden hairs when young; **blade** 5-10 cm long, triangular, sharply pointed, clothed with thick purple or black felt of bristly hair, hardly auricled; **ligule** narrow, entire, or fringed with whitish hairs. **Leaves** 20 x 2.5 cm, often much smaller, linear or linear lance-shaped; margins rough; tip sharp, stiff, main nerves 4-6, with transparent or translucent glands, base rounded, petiole 0.3 cm. often swollen. **Inflorescence** an enormous panicle, often comprising the whole culm. **Spikelets** 3, small, fimbriate on the edges. Stamens 6, slender drooping. Ovary elliptic with 3 feathery stigmas. **Grain** ending in a short beak. **Distribution:** South India, Sri Lanka and Myanmar; widely cultivated in North India up to 1000 m. **Phenology: Flowering** gregariously at intervals of 32-34 years. It flowered throughout India in 1970-71. After flowering it dies down, reproducing itself from seed. **Miscellaneous:** It makes a close impenetrable tall fence. In Hyder Ali's reign, Bednore in Mysore was defended by a deep trench with clumps of the thorny bamboo. Extensively used for building, ladders, tent poles, split stems are woven into mats, baskets and in the manufacture of good quality paper.

130. TIGER BAMBOO, YELLOW OR GREEN STRIPED BAMBOO, GOLDEN BAMBOO
Bambusa vulgaris var. *vittata* A&C Riviere

(Family: Bambuseae)

Plate 12

Chittagong *Briala*; Bengali *Basini Bans*; Marathi *Vansa kulluka*; Tamil *Ponmoongil*; Sri Lanka *Una*.

Field Identification: Culms yellow, striped with green, shining, 6-15 m x 5-10 cm in diameter. The only yellow bamboo striped with green. This character alone enables quick identification. **Description:** An ornamental bamboo with distant culms. **Culms** yellow or striped with green, shining, 6-15 m long, 5-10 cm in diameter. **Nodes** narrowly ringed, slightly raised. **Internodes** 25-45 cm long, thin walled. **Culm sheaths** 15-25 x 17.5-22.5 cm, clothed with appressed brown hair, blade triangular, pointed, 5-15 x 10 cm broad, appressed hairy, often striped like the culms when young, auricles round, fringed by bristles; **ligule** 0.5-0.8 cm broad, toothed. **Leaves** linear lance-shaped, pointed pale green 15-25 x 1.8-4.3 cm, petiole 6.5 cm. **Spikelets** 1.5- 2 cm, oblong, compressed, 6-10 flowered, **Lodicules** 3, stamens 6, protruding, anthers purple, pointed. Ovary oblong, hairy with 3 feathery stigmas. **Caryopsis** not known. **Distribution:** Country of origin is not known. *Index Kewensis* mentions Mexico as its native habitat. It is now cultivated worldwide in the tropics. Common in the Subcontinent. **Phenology: Flowering**: It is known for its constant sterility. According to some authors it has not been seen in flower since its description in 1810. (Suri & Chauhan *Indian Timbers*, Bamboo Information Series 28, p.11 (1984), Forest Res. Inst., Dehra Dun); Gamble (1896) in *Bambuseae* of Brit. India p. 44 says 'Obtained in flower by Thwaites (Ceylon) in 1863'. **Propagation** by layers and cuttings. **Miscellaneous:** Very ornamental, popular in parks. It is commonly employed for making flower vases. *B. vulgaris* var *striata* is an ornamental variety. **Etymology:** The popular name tiger bamboo has been coined because of the yellowish striped culms.

131. MALE BAMBOO, SOLID BAMBOO
Dendrocalamus strictus (Roxb.) Nees

(Family: Bambuseae)

Hindi *Bans*; Bengali *Karail*; Marathi *Bhariyel*; Gujarati *Nakor Bans*; Telugu *Sadanapa Vedura*; Tamil, Malayalam *Kal Mungil*; Kannada *Kiri Bidiru*; Oriya *Salia Bhanso*; Burmese *Myinwa*.

Field Identification: Culms densely packed, 7-13 m x 2.5-7.5 cm in diameter, greyish green. Internodes 30-45 cm. Culm sheaths 7.5-30 cm long, golden brown stiff hair beneath, blade triangular, hairy, slightly auricled, ligule narrow. **Description:** The stems are closely packed in dense clumps. Culms solid or with a small cavity, 7-13 m by 2.5-7.5 cm in diameter, greyish-green often blotched. Nodes swollen. **Internodes** 30-45 cm long; upper branches curved, drooping. **Culm sheaths** variable 7.5-30 cm long, covered beneath with golden brown stiff hairs, smooth in dry areas, slightly auricled; blade triangular, hairy; ligule narrow. **Leaves** deciduous, narrow lance-shaped, 10-25 cm long, finely hairy on both sides. **Inflorescence** a large branching panicle of dense globose heads 2.5-3.7 cm across. **Spikelets** spinescent, flowering glumes ovate, ending in a sharp spine surrounded by tufts of hair. Stamens 6, long, protruding, anthers yellow with a pointed tip. Ovary top-shaped ending in a purple feathery stigma. **Caryopsis** brown and shining 0.7 cm long. **Distribution:** Throughout the Subcontinent including Myanmar in deciduous forests except North Bengal, Sikkim, Bhutan, Assam, Arunachal Pradesh and moist areas of Western Ghats. Myanmar. Cultivated in the plains and foothills. **Phenology: Leaves** fall in February-March. New leaves appear in April. **Flowers** Sporadic flowering every year from November-April. **Seeding**: Ripe seeds fall from April-June. The period between 2 gregarious flowerings over the same area is 30-45 years.

Miscellaneous: Used for batons, 'lathis', building purposes, bows and arrows, tent poles, paper pulp, leaves as fodder. This is the most widely used bamboo of the Subcontinent. It is popularly called the male bamboo because of its strength and due to culms being solid.

132. RHINO BAMBOO
Dendrocalamus hamiltonii Nees Arn. ex Munro

(Family: Bambuseae)

Hindi *Kaghzi Bans*; Bengali *Pecha*; Lepcha *Pao*; Nepali *Tama*; Bhutan *Pashing*; Assamese *Kokwa*; Burmese *Wabo-myetsangye*.

Field Identification: Rhizome shaped like a rhinoceros horn. Culms to 26 m by 12 cm in diameter, soft. Internodes 30-50 cm. Culm sheaths triangular, persistent, to 45 cm long, with brown patches of hair, ligule 0.45 cm broad, auricle acute, blade ovate lance-shaped. **Description:** Rhizome shaped like a rhinoceros horn. A large tufted bamboo. **Culms** erect, often slanting, or curved, 26 m tall, 12 cm in diameter, thin walled and soft and hence the Hindi name 'Kaghzi bans'. Culms greyish white with dense appressed hair when young, dull grey when old. **Internodes** 30-50 cm long. **Culm sheaths** triangular, persistent, 40-45 cm long, with patches of brown appressed hair; auricle acute, blade ovate lance-shaped with black hair and incurved margin; ligule 0.45 cm, smooth. Branches on lower parts of stem seated on woody knobs the size of a fist and on these may be swollen buds with brown sheaths. **Leaves** 37.5 x 6.2 cm broadly lance-shaped. Leaf sheaths with white appressed hairs. **Spikelets** in dense rounded clusters 1.5-4 cm across, purple. Stamens 6, protruding out, pendant, anthers purple, connective with a black hairy twisted tip. Ovary with a bifid feathery stigma. **Caryopsis** ovoid. **Distribution:** Central and Eastern Himalaya, Assam, Garo, Khasi and Naga Hills and Myanmar to 1000 m. **Phenology:** It flowers both annually and gregariously, the latter after 30 years. **Flowers** November through March. **Seeds** ripen April-May. **Miscellaneous:** For making baskets, mats and screens, being soft it is not suitable for building purposes. Young shoots are eaten as a vegetable. Garos cook rice in the internode over a fire and keep turning it until it is charred and then split open the bamboo to take out the rice. The rhizome which after cutting is a replica of the rhino horn is indistinguishable by ordinary means from the real horn and fetches a fabulous price for smugglers. **Etymology:** The popular name rhino bamboo has been coined because its rhizome is an exact replica of a rhinoceros horn.

133. GIANT BAMBOO
Dendrocalamus giganteus Munro

(Family: Bambuseae)

Burmese *Wabo*; Assamese *Worra*.

Field Identification: At once recognized by its giant size 26-33 m tall, by 20-30 cm wide and by large sheaths 50 x 50 cm. **Description:** A giant evergreen bamboo with closely packed culms. **Culms** 26-33 m tall by 20-30 cm in diameter, rarely up to 37 m tall, erect, thin walled, grey green covered with white waxy scurf when young. **Internodes** 37.5-40 cm long; nodes hairy beneath marked with root scars. **Culm sheaths** hard, shining, very large 50 cm long by 50 cm broad at the base, falling off early, thinly covered with stiff golden hairs, dull yellow; imperfect blade 8-37 cm long x 9 cm narrowed upwards into a sharp point; auricles wavy; **ligule** dark brown to black 5-12.5 cm long x 0.6-1.5 cm broad, serrate on the edges. **Leaves** variable in size 22.5-50 x 5-10 cm, oblong, abruptly pointed, unequal sided, glabrous above, hairy beneath when young, edges rough, main vein prominent, secondaries 12-16 pairs, pellucid dotted; **leaf sheaths** striate, glabrous, ending in a callus which is sometimes shining and prominent; **ligule** broad, fimbriate, sometimes cleft in the middle. **Spikelets** 1.2 cm long, ovate, 2-5 in a head, on long spikes, often leafy at top, glumes mucronate, 1-2 empty, 3-6 fertile, stamens 6 with long filaments, anthers 0.75-1 cm long, pointed. Ovary ovoid, with a feathery purple stigma. **Caryopsis** oblong 0.75 cm. A very fast growing bamboo, up to a metre a day has been recorded. **Distribution:** Probably indigenous to Martaban (Myanmar). Cultivated in Assam, Bengal, Dehra Dun, South India and Sri Lanka. **Phenology:** Evergreen. **Flowering**: Occasionally found in flower in Myanmar. In Tenasserim in 1892, South Shan States in 1893 and in Botanic Garden, Calcutta, in 1860-61 and 1883. **Miscellaneous:** In buildings, masts of boats, internodes are made into buckets and flower pots, as petrol containers during World War II.

134. MULI BAMBOO
Melocanna baccifera (Roxb.) Kurz

(Family: Bambuseae)

Syn. *M. bambusoides* Trin

Bengali *Muli*; Assamese *Tarai, Wati*; Burmese *Tabinwa*.

Field Identification: Culms single, distant, 60 cm apart, bright green 8-15 m by 3.7-6.2 cm in diameter. Culm sheaths 12.5-17.5 cm long with a very long narrow tapering blade 30 x 2.5 cm, fine tipped. **Description:** Evergreen, gregarious with single distant culms *c.* 60 cm apart, arising from a long creeping rhizome. **Culms** straight, 8-21 m x 3.7-6.2 cm in diameter, green when young. Nodes marked by a thin ring. Internodes 30-50 cm long, walls thin 0.5-0.7 cm. **Culm sheaths** persistent, yellowish brown, marked with thin furrows, with whitish appressed hair, contrasting with the bright green internodes, 12.5-17.5 cm long, blade very long and narrow 30 x 2.5 cm, tapering to a fine tip, **auricles** round and fringed, **ligule** very narrow, toothed like a saw. **Leaves** lance-shaped or oblong-lance-shaped, long pointed 15-35 x 2.5-8.4 cm, petiole 0.5-1.2 cm. **Inflorescence** a panicle of one-sided, drooping, spicate branches with clusters of 3 to 4 spikelets. **Spikelets** 1.2 cm, spinous. Stamens 5-7, anthers yellow, apex notched. Ovary ovoid, with 4 hairy recurved stigmas. **Fruit:** Large, fleshy, pear-shaped, 7.5-12.5 x 5-7.5 cm, stalked at the thick end, apex terminating in a curved beak. **Distribution:** Garo and Khasi Hills, Mizoram, Bangladesh and Myanmar. **Phenology: Flowers** gregariously over large stretches, at intervals of about 30 years (Gamble, *Bambuseae*, p.120, 1896). Flower buds visible by September-October. **Flowering** December-January. **Fruit** ripens April-June, germinating on the culm itself, 8-10 fruits are seen hanging in clusters round each node. **Miscellaneous:** Fruits are readily devoured by elephants, gaur, rhinoceros, deer, pigs and cattle. Thin walled but strong, extensively used in buildings, house walls, matting, baskets and thatching.

135. CROZIER CYCAS
Cycas circinnalis Linn.

(Family: Cycadaceae)

Hindi *Jangli-madan-must-ka-phul*; Telugu *Kamkshi*; Tamil *Canningav*; Malayalam *Indapana*; Oriya *Oruguna*.

Fig. 70

Field Identification: Evergreen palm-like tree with a cylindrical trunk crowned by glossy fern-like pinnate leaves 1.5-2.5 m long; leaflets 80-100 pairs, margins strongly revolute. Male cone about 33 cm x 7.6 cm. Carpophylls 15-20 cm tapering to a long acuminate point, strongly toothed lobate bearing 2 pairs of ovules. Seeds globose 3.8 cm in diameter, orange red. **Description:** An evergreen palm-like tree up to 8 m high with a cylindrical trunk and a crown of glossy fern-like, stiff but gracefully curved and pinnate leaves; trunk clothed with compacted woody bases of petioles, usually columnar, simple, but often branching when the terminal bud has been cut off. **Leaves** pinnate 1.5-2.5 m long, petiole 0.4-0.6 m long, with short distant spines at right angles to the petiole. Leaflets 80-100 pairs, alternate 15-30 cm x 7-12 mm, blunt or acute, margins strongly revolute. Male **cone** about 33 cm long and 7.6 cm in diameter; antheriferous scales long acuminate, acumen

Fig. 70

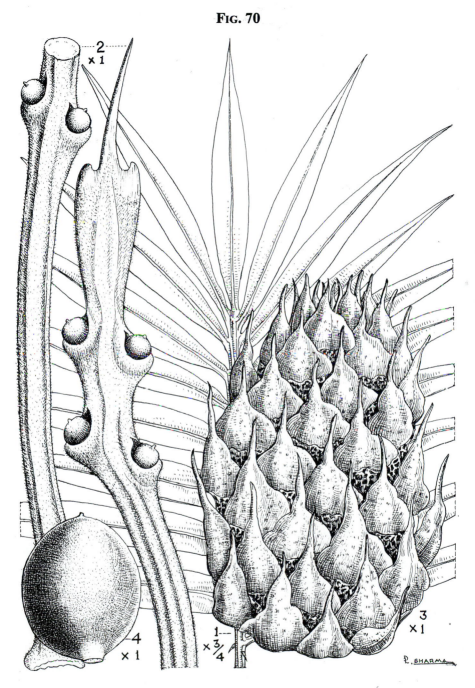

S.No. 135. *Cycas circinnalis*. 1) Leaves x $\frac{3}{4}$. 2) Carpophylls with two pairs of ovules x 1. 3) Male cone x 1. 4) Seed x 1. (From *Khant* 1323, Burma, *Curtis,* 2828, 2827).

in upper half of cone strongly reflexed, erect in the lower half. **Carpophylls** 15-20 cm long, blade 7.6 x 2.5 cm broad, ovate lanceolate, tapering into a long acuminate point, strongly toothed, lobate, bearing two pairs of ovules. **Seed** glabrous, 3.8 cm in diameter, orange red. **Distribution:** Dry deciduous forests of Southern India and Sri Lanka from sea level to 1,070 m and also in the hill forests of Puri and Angul in areas of heavy rainfall. **Phenology: Flowers** February-March. **Fruit** ripens August-October. **Miscellaneous:** The leaves are used for making mats, young shoots are edible. Sago is obtained from the trunk. For obtaining Sago the trees are felled at the age of seven years, the trunk cut up into disks and pounded into flour. Seeds also yield starch. *C. circinnalis* Linn. var. *orixenensis* occurs in the hill forests of Puri.

136. YEW
Taxus baccata Linn.

(Family: Taxaceae)

Hindi *Thuner*; Kashmiri *Birmi, Postil*; Bengali *Bhirmil*; Lepcha *Cheongbu*; Nepal *Tcheragulab*; Khasi *Barmi, Birni*.

Field Identification: Small evergreen tree with reddish-brown thin bark. Leaves linear, flattened 2.5-3.8 cm long. The female cones are one-seeded and surrounded at the base when ripe by coloured fleshy covering. **Description:** A small evergreen tree up to 6 m and 1.5-1.8 m girth with a fluted stem with reddish-brown thin bark. Branches spreading, not whorled. **Leaves** 2.5-3.8 cm long, linear, flattened, arranged in 2 opposite vertical rows, dark glossy green above, paler beneath. **Flowers** usually dioecious, (rarely on the same tree). Male flowers in catkins which are sub-globose and solitary in the leaf-axils, stamens about ten, pollen sacs 5-9, globose, arranged around the filament beneath the tip of the stamen. Female flowers solitary, axillary, each consisting of a few imbricate scales around an erect ovule, which is surrounded at the base by a membranous cup shaped disk. In fruit the disk enlarges, becomes succulent and bright-red, 8 mm long, and surrounds the olive green seed. **Distribution:** All along the Himalaya, Meghalaya, Naga Hills, Manipur and in Myanmar above 1,800 m in the evergreen and coniferous forests. Slow growing and long lived tree, up to 2000 years. **Phenology:** The needles are shed in May and June. Flowers March-May and the seeds ripen from September to November of the same year. **Miscellaneous:** In Ladakh the bark is used in place of tea. The timber is known for its durability and strength and is used for ploughs and axles of carts. The seeds are poisonous to humans. Himalayan yew, a source of the anti-cancer drug Taxol, is now proved to be a richer source of Taxol precursors than the American yew.

137. LARGE LEAVED PODOCARP
Podocarpus wallichianus C. Presl.

(Family: Podocarpaceae)

Syn. *P. latifolia* Wallich

Tamil *Narambali*; Kannada *Kurunthumbi*; Khasi *Soplong*; Lushai *Thingromao*; Burmese *Thitmin-ma*.

Field Identification: Evergreen tree with large ovate lanceolate leaves 8.8-19 x 2.8-6.3 cm. opposite to sub-opposite. Male spikes 1.27 cm long surmounting a stalk 1.25-1.85 cm long. Seed solitary 1.9 cm across, blue-black, seated on a fleshy receptacle. **Description:** Evergreen tree up to 24 m high. Bark smooth, mottled brown and white, wood aromatic. **Leaves** opposite or sub-opposite, 8.8-19 x 2.8-6.3 cm; ovate, narrowing at the apex to a long fine point. **Male spikes** 1.27 cm long, 2-5 together, sessile or on a short peduncle. **Seed** ovoid 1.9 cm blue-black. Seated on a fleshy receptacle. **Distribution:** This is the only naturally occurring conifer in South India. It occurs in Assam, Khasi Hills, Martaban and Tenasserim in Myanmar and in the Western Ghats from the Nilgiris southwards, usually at altitudes of 900-1500 m. It was collected for the first time from the Gt. Nicobar Island in 1953 by the author. **Phenology:** Flowers rainy season. **Fruits** cold season.

138. CHIR PINE, HIMALAYAN LONG-NEEDLE PINE
Pinus roxburghii Sargent

(Family: Pinaceae)

Syn. *P. longifolia* Roxb.

Plate 13

Hindi *Chir, Chil*.

Field Identification: A large tree, female cones solitary or 2-5 together, 10.2-20 x 7.5-10 cm, with thick woody scales, scales pyramid-shaped with a pointed or curved beak. Seeds 7.5-10 mm with a long thin membranous wing. Needles in bundles of three, 20-33 cm long. **Description:** A large evergreen tree, nearly deciduous in dry localities, usually not exceeding 30 m in height, rarely 55 m. A girth of 3 m has been recorded under favourable conditions. Branches up to middle age whorled, crown elongated to pyramidal shaped afterwards becoming spreading or umbrella shaped. **Bark** 2.5-5 cm thick, exfoliates in long plates, brownish-red, turning light grey. **Leaves** in bundles of 3, 20-33 cm long, bright green, each bundle covered at the base by a persistent sheath. **Flowers** monoecious. Male catkins 1.3 cm long. **Female cones** solitary or 2-5 together. Cones 10.2-20 x 7.5-13 cm, with thick woody scales, the scales are pyramid-shaped, with a pointed or curved beak. **Seed** 7.5 mm-1 cm long with a long thin membranous wing. **Distribution:** All along the Himalaya from Pakistan to Arunachal Pradesh at 450-2300 m in the outer ranges, where the full force of the monsoon is felt. In the Kashmir valley proper, where the monsoon is scanty because of the elevated Pir Pinjal, it is absent. Gregarious, forming pure forests. **Phenology:** February-March, the needles persist for about one year and five months. They are shed in May-June. **Flowers:** February-April. Seeds are shed in April-May. At low elevations the male inflorescence is visible in early January. The flowers ripen, and pollen showers occur from February-April according to altitude. The reproduction of conifers is best exemplified by chir pine. The male inflorescence is visible in January. As it matures, the scales open, showering in abundance sulphur dust-like pollen grains. Each pollen grain has two balloon-like wings which aid dispersal by wind and alight on the female cones. A year elapses between pollination in the spring and fertilization. At the time of pollination the scales of the female cones open to receive the pollen. The female cones grow fast to adult size after fertilization, the scales thickening, bending back and closing tightly. They mature in 25 months and open about May in the dry weather to disgorge the winged seeds. The mature seeds bearing wings are carried far and wide by the wind. Seeds falling on the ground germinate when the monsoon breaks. Some are eaten while on the tree by flying squirrels. **Miscellaneous:** It produces resin of commercial importance. Useful joinery timber and used as railway sleepers after treatment.

139. CHILGOZA PINE
Pinus gerardiana Wall. ex Lamb.

(Family: Pinaceae)

Plate 14

Hindi *Chilgoza, Neoza* (seeds); N.W. Himalaya *Chiri, Gunober Prita.*

Field Identification: Needles in threes, stiff 5-10 cm long, dark-green. Female cones oblong-ovoid 15-23 x 10-13 cm. Scales 3.8 x 2.5 cm, very thick and woody with a stout recurved beak. Seed cylindric, 2.5 cm long, oily. **Description:** An evergreen tree 15-18 m, occasionally up to 24 m high, usually 1.8-2.4 m in girth. **Bark** silver grey, thin, smooth, peeling off in large thin flakes. **Leaves** in threes, 5-10 cm long, stiff, dark-green. **Male** catkins 7.6-13 mm long. **Female cones** near the middle of the shoot. Mature cones oblong-ovoid 15-23 x 1-13 cm. Scales 3.8 cm long, 2.5 cm broad, very thick and woody with a stout recurved beak. **Seed** 2.5 cm long, cylindric, oily with a short caducous wing. **Distribution:** NW.

Himalaya in areas with scarce rainfall but heavy winter snowfall. Unlike the chir, the chilgoza is not found in places with a monsoon rainfall. Local in the inner semi-arid valleys from Bashahr westwards to Kashmir, Chitral, Northern Baluchistan and Afghanistan between 1,800-3,000 m. **Phenology: Leaves** evergreen, which fall by the second year. New leaves in April-May. The **flowers** appear in May-June, when pollination takes place. The young female cones increase in size rapidly during the second year, ripening about September-October of the year after. The seeds are not viable for long. **Miscellaneous:** The seeds are the chief economic product, for they constitute a staple article of food in the W. Himalaya and are exported to the plains of the Subcontinent. The value of the seeds prevents any exploitation of the timber. **Etymology:** It was discovered by Captain Gerard after whom it was named and introduced into cultivation by Lord Auckland in 1839.

140. KHASI PINE
Pinus kesiya Royle ex Gord.

(Family: Pinaceae)

Syn. *P. insularis* Endlicher

Khasi *Ding-se, Dieng-kvsi*; Lushai *Far*; Bengali *Saral*; Burmese *Tinvu, Tinshu*.

Field Identification: Evergreen gregarious tree with a rounded crown. Leaves in threes, 15-25.5 cm, slender, apex a long fine point. Cones solitary or in pairs or occasionally in threes. Ovoid, symmetrical, bright brown 5-7.5 x 5 cm scales woody, thickened towards the apex, with a blunt knob at the end. Seeds about 8 mm long, wing four times the length of the seed. **Description:** A small or moderate-sized gregarious tree with a rounded crown, a smaller tree in Khasi Hills, but attaining 30-45 m height in Myanmar. Branches whorled, **bark** thick, deeply and reticulately cracked, coming off in thick small plates. **Leaves** in bundles of three, 15-25.5 cm long, slender, apex a long fine point. **Cones** solitary, in pairs or occasionally in threes, ovoid, symmetrical, bright brown, 5-7.5 cm long about 5 cm across, stalks short and stiff, scales of cone in spirals of 8-5, woody, thickened towards the apex, with a blunt knob at the end. **Seeds** about 8 mm long with a round-topped wing, which is four times the length of seed. **Distribution:** Khasi and Naga Hills, Manipur, Upper Myanmar and the Philippines. It occurs at elevations of 800-2,000 m in Khasi hills where it thrives at elevations of 1,200-1,400 m. **Phenology:** The new **leaves** appear in February-May. The old needles fall for the most part in April-May, by which time the new ones are full sized. The needles persist for 1 year and 1-3 months. The male and female **flowers** appear in

February-March. The female flowers, 1-3 at the apices, are about 1-3 x 0.9-1 cm in diameter, light green, ovoid. By next year the cones are nearly full sized but still green; they ripen from February-April, two years after the appearance of the female flowers and persist long on the tree. **Miscellaneous:** The quality of resin and turpentine is valuable. It is a fast growing pine and is in demand for afforestation in subtropical countries of the world as well as on the hills of South India.

141. BLUE PINE
Pinus wallichiana A.B. Jackson

(Family: Pinaceae)

Syn. *P. excelsa* Wall. ex D. Don.

Hindi *Kail*; Kashmiri *Kail, Yiro*; Himachal *Lim*; Kumaon *Raisalla, Lamshing*; Lepcha *Neet-Kung*; Bhutan *Tongschi, Lamshing*.

Field Identification: A large tree, female cones 2-3 together, banana-shaped 15-30 cm long, drooping. Needles in drooping bundles of five, with a bluish green tinge for which it is known as blue pine. **Description:** A large evergreen tree, up to 50 m tall. **Bark** greyish brown with shallow fissures. **Leaves** needle-like in bundles of five, 10-20 cm long, drooping, with a bluish-green tinge. The male and the female **flowers** are borne on the same tree. The male flowers (catkins) are 1-2 cm long, ovoid or oblong and are borne on the current year's young shoot forming a cluster about 5 cm long. The catkins fall after ripening. The young female cones are erect initially. On ripening they become pendulous, 2-3 together, banana-shaped, 15-30 cm long, light brown with thin scales, closely adhering to the cone axis like the scales of a fish. **Seed** bluish, egg-shaped, compressed, 6.5- 9 x 3.5-5 mm with a wing about 3 times as long. **Distribution:** All along the Himalaya from Pakistan to Arunachal Pradesh through Bhutan, very rare in Sikkim and a considerable portion of Kumaon 1,800-3,700 m. **Phenology:** New leaves March-April, attain full size by August-September and shed the next year from May to July. **Male flowers** ripen and shed their pollen from April to June and fall on the ground. The **female flowers** are visible in April, pollinated from April to June, with scales open. After pollination the scales close. The **cones** ripen about September-November of the second year. The winged **seeds** are carried far and wide when the cones open during the dry weather in October-November. A crop of cones may be destroyed by monkeys and flying squirrels. The Himalayan nutcracker, a brown bird spotted with white is a dweller in blue pine and fir forests along with the white-browed blue flycatcher. **Miscellaneous:** Railway sleepers, house construction and light furniture. The wood is superior to chir pine. Sometimes a sweet liquid known as honey-dew is secreted by aphids infesting the leaves and is eaten by the Himalayan people.

142. ARUNACHAL PINE
Pinus bhutanica Grierson, Long & Page

(Family: Pinaceae)

A five-needle pine allied to *P. wallichiana*, it differs in having longer needles 12-28 cm and a narrower tapering crown with pendulous branches. In *P. wallichiana* the crown is wider with ascending branches and bigger seeds, 6-10 mm and wing 1.5-3 cm. In the former the seed is 6-8 mm and wing 2 cm. Needles mostly 11-18 cm in the blue pine. Ventral needle resin canals in *P. bhutanica* 1-2 marginal or submarginal, asymmetric. In blue pine 0-1, median symmetric. Dorsal resin canals are similar in both these pines. They are two and marginal. **Miscellaneous:** Discovered by K.C. Sahni in April 1977 from Tenga valley, Arunachal Pradesh and incorporated in the Forest Research Institute Herbarium, Dehra Dun (Sahni & Naithani Ser II 381). In 1979 it was also found by Grierson & Long from Bhutan and described by them as a new species along with C.N. Page at Royal Botanic Garden, Edinburgh. As it was first collected from Arunachal, I have given it the appropriate popular name 'Arunachal Pine'. Sahni & Naithani, No. 381 was sent to Dr. Page for comparison with their known Chinese five-needle pine collections. Grierson *et al.* loc cit p. 299 commented "Independently .... Sahni in April 1977 recognised a pine in Tenga with long needles, but winged seeds and cones, the identity of which he rightly questioned, noting its differences in needle anatomy from *P. wallichiana*."

143. MERKUS PINE, TENASSERIM PINE
Pinus merkusii Jungh. et De Vr.

(Family: Pinaceae)

Burmese *Tinvu, Tinshu, Shaja.*

Field Identification: Only two-needle pine of the Subcontinent found in Myanmar. Cones usually in pairs, 5-7.5 cm long, cylindric conical, erect, seeds many times shorter than the wing. **Description:** A moderate sized to large tree attaining 18 m in Myanmar and 30 m in Sumatra, of conical habit when young; spreading or round headed when mature. **Leaves** in bundles of two, 15-25 cm. **Cones** usually in pairs, 5-7.5 cm long, cylindric-conical, erect, peduncle 1.3 cm long, scales with a thick but flat pyramidal beak, faces of which are grooved or furrowed lengthwise. **Seeds** small many times shorter than the unequal sided wing. **Distribution:** In Myanmar it occurs on low hills and spurs at elevations of 150-760 m associated with *Dipterocarpus tuberculatus*. It extends from the southern part of the Southern Shan States southwards through the hills of the Salween and Thaungin drainage, also found in Thailand, Cochinchina, Sumatra, Java, Borneo and the Philippines. The interesting feature of this species is that it is the only pine in the world that crosses the equator. **Phenology:** New leaves appear in February-March, old leaves fall early in the second year. Flowers February-March, cones ripen April-May. **Miscellaneous:** General purpose construction timber for indoor work. Produces resin of good quality.

144. WEST HIMALAYAN FIR, WEST HIMALAYAN LOW-LEVEL SILVER FIR
Abies pindrow Royle

(Family: Pinaceae)

Kashmiri *Budlu, Badar, Tung*; Garhwal *Morinda*; Kumaon *Span, Krok-Kunawar, Righa.*

Field Identification: A large tree with dense conical columnar crown. Leaves dark green 2.5-5 cm two ranked, on opposite sides of a twig, needle-like, flattened. Female cones erect, cylindric 10-18 x 4-7.5 cm, very dark purple. Seed 8-13 mm, brown shining, wing longer than the seeds. Cones break up on the tree. **Description:** A lofty evergreen tree with a dense conical columnar crown of dark green foliage, attaining a height of 45-60 m and a girth of 2.5-4 m. Height of 63 m and girth of 6 m have been recorded. **Bark** smooth on young stems, dark-grey, rough with deep vertical furrows on old stems. **Leaves** variable, 2.5-5 cm long, two-ranked, on opposite sides of a twig; needle-like, flattened, lower surface with two pale powdery bands on either side of the midrib, tip notched. **Flowers** monoecious. Male catkins 1.3-1.8 cm long, clustered; stamens with 2 linear pollen sacs. Female flowers in cones, which are solitary or in distant pairs, erect, dark purple. **Cones** erect, cylindric 10-18 x 4-7.5 cm. Seed 8-13 mm long, brown shining, wing longer than the seed. **Distribution:** Afghanistan to Nepal throughout the western Himalaya at 2,300-3,350 m, sometimes descending below 2,150 m. **Phenology:** The **needles** persist for 3-6 years. The old needles fall chiefly in May-June. The new shoots appear in April-May. **Flowers** April, cones ripen in September and break up on the tree, shedding the seed in October-November. **Miscellaneous:** A useful lightwood, used in making fruit crates, planking for ceiling and floor boards, shingles, camp furniture, etc. It is in demand for paper making. Treated sleepers are used by Railways. An allied species *A. spectabilis* (D. Don) Spach is called the high level fir, found at 2,750-3,950 m rarely up to 4,250 m. This fir is distinguished from *A. pindrow* in having a wider crown, slightly upturned branches, shorter thicker leaves 1.3-2.5 cm, cones shorter and thicker about 7 cm long. The Monal Pheasant is a dweller of fir, birch and rhododendron forests from 2,600-5,000 m.

145. EAST HIMALAYAN FIR
Abies densa Griff.

(Family: Pinaceae)

Plate 15

Nepal *Gobre Salla*; Bhutan *Dunshing*.

A pagoda-shaped tree with horizontally spreading branches, distributed from Nepal to Arunachal Pradesh up to the NE frontier of Myanmar at 2,750-3,950 m. Above 3,200 m it forms pure forests. Young shoots reddish-brown. **Leaves** 2.5-6 cm with recurved margins and silvery undersurface. **Flowers** monoecious. Male catkins clustered. Female **cones** erect, scales thin, breaking away from a woody axis when ripe, fan-shaped contracted by a short claw. Ripe cones cylindric, thicker and shorter than *A. pindrow*. Cones in all Himalayan firs are a deep purple-black. **Phenology: Flowers** April; **Cones** ripen in September-October.

146. INDIAN HEMLOCK FIR, HIMALAYAN HEMLOCK
Tsuga dumosa (D. Don) Eichler

(Family: Pinaceae)

Syn. *T. brunoniana* Carr.

Nepal *Changathasi, Thingia, Sula*; Bhutia *Tanshing*; Lepcha *Semedung, Chemdung*.

Field Identification: Tall evergreen tree, like fir in foliage but at once distinguished by much smaller globular female cones, 1.3-2.5 cm long. **Description:** A tall pyramidal tree up to 36 m high with a girth of 8.5 m. Trees up to 51.8 m in height have been recorded. **Bark** thick and rough, branches gracefully drooping. **Leaves** 1.27-2.54 cm long, more or less 2 ranked, on opposite sides of twig, needle-like, lower surface silvery white. **Flowers** monoecious. Male catkins 0.25-0.5 cm long. Stamens with 2 rounded pollen sacs. Ripe female cones 1.3-2.5 cm long, ovoid, scales rounded shining, leathery, persistent. **Seeds** 0.78-0.88 cm long, wing oblong, obtuse. Hemlock is like fir in foliage, but is easily identified by much smaller female cones (1.3-2.5 cm) in the former. Due to its resemblance with fir it is also called Indian hemlock fir. **Distribution:** From E. Kumaon through Nepal, Sikkim, Bhutan to Arunachal Pradesh and Upper Myanmar. Chiefly at 2,400-3,000 m. **Phenology: Flowers** May-June. The cones ripen the same year. **Miscellaneous:** In Sikkim the timber is made into shingles, and the thick rough bark is used for roofing.

147. WEST HIMALAYAN SPRUCE
Picea smithiana (Wall.) Boiss.

(Family: Pinaceae)

Plate 16

NW. Himalaya *Rai, Rau, Riar, Kachlu*; Garhwal, Kumaon *Roi Ragha, Rhai, Morinda.*

Field Identification: Large tree with pendulous branches. Leaves 2.5-3.8 cm, needle-like with sharp points, growing spirally around the twigs. Female cones pendulous, solitary 10-15 cm x 2.5-5.08 cm, dark brown. Seed 0.5-0.6 cm, dark-grey to blackish, wing spoon-shaped, large seed with wing 1.27 cm long. **Description:** A very large evergreen tree of pendulous habit attaining a height of 60 m and above and a girth up to 5.7 m; branches whorled, horizontal, the branchlets hanging like tassels. **Bark** rough, cut by shallow furrows into small plates. **Leaves** 2.5-3.8 cm long, needle-like with sharp points, arising singly and spirally around the branches, 4 sided and stiff. **Flowers** monoecious. Male catkins 2.54 cm long, solitary, erect. Female flowers in **cones** which are solitary, terminal 10-15 x 2.54-5.00 cm, pendulous, dark-brown. **Seed** 0.5-0.6 cm dark-grey or blackish, wing spoon-shaped total length of seed with wing 1.27 x 0.7 cm; wing light brown. **Distribution:** In the Himalaya from Afghanistan to Kumaon at 2,150-3,300 m, commonly mixed with fir, deodar, blue pine and at higher elevations with *Quercus semecarpifolia*, maple, etc. **Phenology:** The **needles** persist for three to six years, shed mostly in May and June and new ones appear in April. The **pollen** is shed from mid April to mid May. Soon after pollination, the scales close and fertilization is completed. By July the **cones** reach full size though they are still green, ripening in the second season. **Miscellaneous:** Used for planking in ceilings and floors, cheap furniture, crates, rosin casks. Bulk of the supply goes to the Railways as treated sleepers. Suitable for glider construction and was in use by aircraft establishments in India. An allied spruce *P. spinulosa* Griff (Henry), E. Himalayan spruce, occurs from Sikkim through Bhutan to Arunachal Pradesh at 2,600-3,300 m and is characterized by flat needles.

148. DEODAR

Cedrus deodara (Roxb. ex Lambert) G. Don

(Family: Pinaceae)

Sanskrit *Devadaru, Deodaru*; Hindi *Deodar*.

Field Identification: Leaves in tufts of 15-20, dark green, 2.5-3.8 cm, sharply pointed. Female cones erect, barrel-shaped 7.6-12.6 x 5.8 cm, scales fan-shaped 3.2-3.8 cm across. Seeds triangular with a broad wing, seed and wing 2.5-3.3 cm long. The female cones break up on the branches. **Description:** An evergreen tree exceeding 60.9 m in height and 10.6 m in girth. The tallest deodar of Manali (Himachal Pradesh) at 76 m is taller than the 73 m high Qutub Minar in Delhi. **Bark** greyish-brown. Leading shoots of young trees and branches pendulous. Branches of old trees horizontal. **Leaves** 2.5-3.8 cm long, dark green, sharply pointed, in tufts of 15-20. Male and female **flowers** often on separate trees, but sometimes on the same tree in which case they are on separate branches. **Cones** erect, barrel-shaped, 7.6-12.6 x 5.8-8 cm. Scales numerous, fan shaped, 3.2-3.8 cm across. **Seeds** triangular with a broad wing, the seed and wing 2.5-3.2 cm long. **Distribution:** From Afghanistan to Garhwal, Kurnauli Valley (W. Nepal) at 1200-3300 m. Commonly between 1,700 and 2,400 m. **Phenology:** New shoots appear in March or early April. The old leaves are shed chiefly in May, but shedding may also take place in autumn about the time the cones ripen. Male **Flowers** appear in August and ripen after pollination in November of the following year i.e. flowering to ripening is about 13 months. **Miscellaneous:** It is the strongest of Indian coniferous woods. Due to the presence of oil, it is durable and rarely attacked by termites. It is in demand for railway sleepers and as constructional timber. The Western Tragopan, a spectacular and endangered bird was localised in Kashmir, Himachal Pradesh and Pakistan to deodar forests of the upper reaches. The species name is derived from the Sanskrit *devadaru*, Tree of the Gods, whence the common Indian name *deodar*.

149. HIMALAYAN LARCH

Larix griffithiana Hort. ex Car.

(Family: Pinaceae)

Nepal *Boarga sella, Biny*; Lepcha *Sah, Saar*.

Field Identification: A tree with long drooping branchlets with light green foliage. Foliage like deodar, but light green, in tufts of 10-19, 2.5-3.2 cm long, ending in a black point keeled below, deciduous in autumn. Cones cylindrical 5-10 x 2.5-3.2 cm coppery red during growth; unlike deodar persistent for a long time, not breaking on the tree. Seeds winged, in pairs in each scale, 1-1.25 cm including the wing. Without wing 0.4 cm. **Description:** A graceful tree up to 18 m with a spreading crown and long pendulous branchlets. **Bark** thick brown. The only deciduous conifer of the Subcontinent. Young shoots downy, reddish-brown the second year. Terminal buds conical, covered with hairy scales; lateral buds egg-shaped, downy. **Leaves** 2.5-3.2 cm long, light green, ending in a black point, deeply keeled below, in tufts of 10 to about 19. Male **flowers** 0.9 cm long on short, stout stalks. Female flowers ovoid, bracts long reflexed. **Cones** cylindrical 5-10 x 2.5-3.2 cm, coppery-red during growth, the short stout stalk twisted to bring the point of the cone upward on the pendant shoot. **Seed** 0.4 cm long, winged, in pairs on each scale 1-1.25 cm long including the wing. **Distribution:** The **needles** are shed in the autumn, the new foliage appearing in the spring. The **flowers** appear in May and the **cones** ripen in October of the same year. **Miscellaneous:** A graceful ornamental tree, particularly when the cones are developing, which are of the colour of 'litchi' fruits. The species is not of importance as a timber tree. In Sikkim it is used for making shingles.

150. HIMALAYAN PENCIL JUNIPER
Juniperus polycarpos C. Koch

(Family: Cupressaceae)

Syn. *J. macropoda* Boiss.

Uttar Pradesh *Dhup, Padam*; Himachal *Chalni, Lewar, Shupka*; Nepal *Dhupi, Chandan*.

Field Identification: A shrub or medium-sized tree 12-15 m tall, with a girth of 1.8-2 m. Crown pyramidal. Leaves pungent, of the lower branches awl-shaped and of the upper scale-like, closely compressed. Fruits subglobose 8.5 mm across, bluish-black when ripe, 2-5 seeded. Wood fragrant. **Description:** A shrub or medium-size evergreen tree 12-15 m or sometimes taller with a trunk of 1.8-2 m. Trees up to 10 m girth and estimated to be 1000 years old, are recorded from Lahaul. **Leaves** of the lower branches awl-shaped, pungent, of the upper branchlets scale-like, closely compressed, egg-shaped, pointed. **Male catkins** on a scaly stalk. **Fruit** subglobose 8.5 mm across, bluish-black when ripe, 2-5 seeded. **Distribution:** Afghanistan, Baluchistan, Kagan Valley, Kashmir, Lahaul extending to Kumaon and W. Tibet from 1500-4300 m, in the inner semi-arid ranges. It forms forests of considerable extent, particularly in Kagan Valley. **Phenology:** Flowers in the spring. Fruit ripens in September-October of the second year. **Miscellaneous:** The wood is eminently suitable for pencil making and also burnt as incense in monasteries.

151. HIMALAYAN CYPRESS
Cupressus torulosa D. Don

(Family: Cupressaceae)

Hindi *Leauri*; W. Himalaya *Devidar, Cella, Surai, Raisal*.

Field Identification: Trees up to 45 m high with drooping branches. Female cones globose 12 to 17 mm across, occurring in clusters, woody. Seeds compressed, 6-8 on each scale, about 5 mm across, pale brown winged. Leaves small 2.5 mm long or less, triangular, thick closely set on the twigs. **Description:** A large evergreen tree commonly 3.6 m in girth and 45 m high, with a pyramidal crown with horizontal or drooping branches. **Bark** peeling off in long thin strips. **Leaves** 2.5 mm long or less, triangular, thick with closely pressed tips so that the branches are not rough to the touch. **Flowers** monoecious, male catkins 5-6 mm long, often tinged purple, solitary at the tips of the branchlets. **Cones** globose, scales 6-10. Ripe cones 12-17 mm in diameter, clustered, scales woody. **Seed** compressed, 6-8 on each scale, about 5 mm across, pale-brownish. **Distribution:** Outer ranges in the Himalaya from Chamba (Himachal Pradesh) to Nepal, Kameng District (Arunachal Pradesh). It also occurs in W. Szechuen China, where it is common in upper semi-arid reaches. Its altitudinal range in the Himalaya is from 1800-2800 m. The trees in Kashmir and Pakistan are apparently cultivated. **Phenology:** Male **flowers** appear in September-October. They ripen and shed their pollen in January or February, the trees become conspicuous with golden yellow covering of male flowers. **Female cones** ripen in the second year. The cones commence opening and shedding their **seeds** from August to December. **Miscellaneous:** Timber is used in making pencils.

152. TREE FERN
Cyathea brunoniana (Wall. ex Hook.) Clarke et Bak.

(Family: Cyatheaceae)

Nepal *Pakpe*; Lepcha *Pashien, Pasen*.

Field Identification: 9-23 m, rachis with prickly sori in 2 rows. **Description:** A large tree fern 9-12 m; main rachis prickly, secondary rachis slightly rough with tubercles or smooth; rachis of pinnules crisped hairy; veinlets forked or three branched; sori in 2 rows close along the main veins; involucre a hemispheric cup. **Distribution:** Nepal, Darjeeling, Sikkim, Bhutan, Khasi Hills, Bangladesh and Myanmar at 1170-2206 m. Also in the monsoon forests of South China across the Himalaya at *c.* 900 m. **Miscellaneous:** Lepchas eat the soft pith.

153. GIANT TREE FERN
Cyathea gigantea (Wall. ex Hook.) Holtt.
= *Alsophila gigantea* Wall. ex Hook.

(Family: Cyatheaceae)

A. glabra of Beddome; Handb. (not *Cyathea glabra* (Bl.) Copel.)

Field Identification: Sometimes over 12 m. Petiole black or very dark. Sori shaped like an inverted V. E. Himalaya, Nilgiris, Sri Lanka. **Description:** A large tree fern. **Stipes** (petiole) black or very dark, slightly rough, scales small, dark with pale thin edges. Rachis (midrib) dark to purplish brown, smooth below, sparsely hairy above. **Middle pinnae** 45 x 18 cm. Pinnules of middle pinnae 14 x 2 cm, nearly sessile, base wedge-shaped, narrowing from the broad base upwards to a pointed apex, lobes deltoid, 5 to 6 mm broad, apex rounded and toothed. **Veins** 5 or 6 pairs in each lobe; scales or costules small, pale, lobed with short hair on the margin, stiff setae at apex of larger scales; sori arranged in the shape of an inverted V; no indusia. **Distribution:** Common in the Eastern Himalaya and Myanmar, Nilgiris and Sri Lanka.

KNOW YOUR TREES

Some of the field and diagnostic characters that aid quick identification are indicated. The writer has drawn freely from works of good field botanists such as R.N. Parker (1918), N.L. Bor (1953) and his own works **Encyclopedia of Indian Natural History,** BNHS (1986). Some of the examples are not treated in this concise book under species account. They are cited below in a few cases to make this chapter more informative. It is suggested that this book be used together with Bor's **Manual of Indian Forest Botany** (*op. cit.*). Only a few outstanding exotics of natural history interest are given, and in such cases B. Hora's **The Oxford Encyclopedia of Trees of the World** (1981) is suggested in order to arouse interest in this fascinating subject.

Shape of trees: Outside the forest tree shapes are often distinctive. Flat-topped trees are the rain tree (*Samanea saman*), *Enterolobium timbouva, Acacia planifrons*, etc. Conical include many Coniferae, also *Mesua ferrea, Polyalthia longifolia*. Round include mango, tamarind, *Pterygota alata*, etc. Trees with weeping branches are weeping willow (*Salix babylonica*), Indian jujube or ber (*Ziziphus mauritiana*), weeping cypress (*Cupressus funebris*), etc. Mature trees leaning towards east include Cook's araucaria (*A. columnaris*).

Branches: Whorled branches are seen in the red silk cotton, *Garcinia*, kadam. The long straight branches drooping at the tips (*Duabanga grandiflora*), drooping branches in mast tree (*Polyalthia longifolia* var. *pendula*), dichotomous branching in branching palm (*Hyphaene dichotoma*), screw pine (*Pandanus* sp.).

Cylindrical boles of Dipterocarps are best seen as one approaches the Andamans by sea. *Calophyllum, Agathis robusta, Poeciloneuron*, etc. Big fat swollen boles of baobab are seen on the west coast.

Buttresses: Enormous plank buttresses are seen in *Tetrameles nudiflora*, a favourite nesting tree of hornbills. Massive buttress roots are seen in tropical rainforests of Andaman and Nicobar Is., Western Ghats, the foothills of the eastern Himalaya, e.g. *Bombax ceiba*, shingle tree (*Acrocarpus fraxinifolius*), one of India's largest trees, a girth of 9 m is on record. *Pterygota alata* has a tripod-like buttress to withstand hurricanes. *Terminalia* sp., *Bruguiera gymnorhiza* which has what are called flying buttresses, meaning trunk supported on stilt roots.

Roots: Stilt roots *Pandanus* sp., *Rhizophora*. In date palm *Phoenix dactylifera* the trunk is surrounded at the base by a dense mass of root suckers, by which it is separated from *P. sylvestris*. The wild date palm is common on the Delhi-Dehra Dun Highway.

Cauliflory: Flowers and fruits are sometimes borne on the trunk. *Gynocardia odorata*, cannonball tree (*Couroupita guianensis*) of South America has cannonball-like fruits 15-20 cm across, borne on the trunk, which drop with a loud crash. Flowers fragrant pink inside orange-yellow outside, 10 cm across. Seen in Botanic Garden, Peradenya, Sri Lanka and elsewhere.

Spines on stems: They are confined to tropical trees. *Bombax ceiba, Chorisia speciosa* recognised by rose-like thorns. *Ceiba, Erythrina, Ziziphus incurva* and *Zanthoxylum* sp., are covered with thick-based squat spines which eventually become cork based and can be struck off with a smart blow. Spines are generally absent in temperate trees. In *Gleditsia* and *Flacourtia* the spines are branched. Some palms have branched spines, as sharp as needles. The sheaths of the leaves of the climbing canes develop needle-like spines. The author recalls with horror the painful pricks he got from needles of palms followed by leech bites while on a hazardous botanical expedition to the Great Nicobar Island in 1952 which, at that time, was *terra incognita*. This island, happily, is now a Biosphere Reserve. The author was rewarded for his pains by having a mountain range named after him *Biosphere Reserves in India*, Ministry of Environment & Forests, New Delhi, June, 1989, p. 59.

Bark: The bark of young trees of pipli (*Exbucklandia populnea*) is black, of older trees brown and deeply fissured. The bark of siris (*Albizia lebbeck*) is black grey. This character, together with its straw coloured pods, at once separates it from white siris (*A. procera*). Boles and twigs of *Diospyros* (ebony family) are black. In lancewood (*Homalium tomentosum*) it is white and this tall tree looks spectacular with its twisted trunk. These trees were introduced into the Forest Research Institute, Dehra Dun from Myanmar. The bark is hard and resonant in the white dhup (*Canarium strictum*). When struck by a blow of a *dao* it gives a ringing sound. The trunk and bark of floss silk tree (*Chorisia speciosa*) first introduced in India by the FRI is greenish. Papery bark in chalta (*Dillenia indica*), bhojpatra (*Betula utilis*), exfoliating in patches or strips in chinar (*Platanus orientalis*), gamari (*Gmelina arborea*) while the sandan has a crocodile-like skin. In *Shorea, Pinus, Calophyllum* and *Erythrina,* it is fissured.

Slash, blaze or cut: The trees below are classified on the basis of the sap or latex colour.

Colourless sap	*Tetrameles nudiflora.*
White latex	Anacardiaceae (*Parishia insignis, Rhus*), Apocynaceae (all species), Asclepiadaceae (all species), Euphorbiaceae (*Euphorbia*), Clusiaceae (*Mesua ferrea*), Meliaceae (*Amoora* spp.), Moraceae (*Morus, Artocarpus*).

White turning muddy brown	Upas tree
Yellow latex	Clusiaceae, most species of *Garcinia, Ochrocarpus, Thespesia* spp.
Red juice	Bixaceae *Bixa orellana*, Euphorbiaceae *Bischofia javanica*, Myristicaceae all species, Papilionaceae *Butea monosperma.*
Red gum	Euphorbiaceae *Macaranga*, Papilionaceae Sandan. Resinous *Canarium*.
Colourless gum	Anacardiaceae *Lannea*, Burseraceae *Bursera*, Dilleniaceae *Dillenia* spp.
White or grey juice	Anacardiaceae *Semecarpus* and other genera. The colour turning black and texture of the slash or blaze is often helpful.

As each one is different, individual trees are known by their blaze. A few examples are given to indicate the diversity of this character.

Lannea coromandelica	Crimson, marked with pink and white.
Schima wallichii	Red, spongy with crystals of oxalate.
Cordia dichotoma	White turning dirty green.
Ougeinia oujeinensis	Blood-red streaks on a white ground.
Celtis australis	Chocolate with lighter specks.
Aromatic blaze	Lauraceae especially *Cinnamomum*.
Blaze with an unpleasant odour	*Celtis* spp.
Blaze smelling of bitter almonds	*Pygeum* spp.
Blaze sweet smelling	*Toona* spp.

Leaves: They develop from buds; the latter are often protected by stipules. In *Dipterocarpus* the stipules are large and the fallen ones are prominent on the forest floor. In Dipterocarpaceae and Magnoliaceae they leave circular scars on the twig after falling. In Clusiaceae and Rhizophoraceae, both with opposite leaves, the 2 stipules grow together and are called interpetiolar stipules, an outstanding character for quick identification.

Leathery leaves with very many parallel nerves can often be placed at once in Clusiaceae (*Calophyllum, Mesua*) or, if covered with rusty scales, it is almost a certainty that they are *Karayani*, habitat of the lion-tailed macaque. Leaves with a prominent intramarginal vein (*Syzygium*, Rhizophoraceae). One of the best features of diagnostic value are pellucid cells in leaves (Rutaceae, Myrtaceae, Hypericaceae). In Samydaceae, streaks and dots (*Casearia*). In *C. glomerata*, the dots turn the colour of dried blood when old. Stellate or star-shaped hair on the under surface (Malvaceae, Sterculiaceae, Tiliaceae). Lepidote or covered with scurfy scales (*Elaeagnus, Cullenia*). Buk oak, *Quercus lamellosa* is recognised by white undersurface of its leaves. Leaves with 3-5 basal nerves (*Ziziphus, Cinnamomum, Strychnos*). Young leaves of gab *Diospyros malabarica* are red. Some leaves are bright blue (*Memecylon*). Some trees have always a few bright red or orange leaves (*Sapium, Elaeocarpus*). In *Terminalia catappa* they turn bright red before falling. The leaves of teak, when rubbed between the hands, colour them red.

Petiole: Peltate petiole that is attached inside the margin of the leaf as seen in *Macaranga*. Winged petiole is seen in some *Citrus*.

Fruits: are distinctive. Pods are characteristic of Leguminosae, but some families develop fruits which are indistinguishable from pods. However, pods of legumes are not always dehiscent, especially in those with only one ovule that develops into a seed, while the rest turns into a wing.

Examples of fruits which are samaroid one seeded pods (*Pterocarpus, Butea, Dalbergia*). Flat pods (*Albizia*). Inflated pods (*Crotalaria*). Cylindrical pods (*Cassia*). Necklace-shaped pods (babul). Horse-shoe shaped (*Enterolobium timbouva*). Sickle-shaped (gulmohur). Round fruits (chalta, *Hydrocarpus, Gynocardia, Pterygota alata*). Capsules with seeds enveloped in floss (*Bombax ceiba, Cochlospermum*). Capsules with winged seeds (*Lophopetalum*). Woody fruits *Lagerstroemia, Schima wallichii*. Lauraceae, Rosaceae, Elaeocarpaceae fruits are very often drupaceous. Follicle i.e. dry dehiscent fruit opening only on the dorsal suture and the product of a simple pistil (Apocynaceae, Bignoniaceae, Sterculiaceae). Some follicles of Bignoniaceae are winged (*Oroxylum, Millingtonia*).

Winged fruits
    Calyx developed as wing or wings: Dipterocarpaceae: nearly all
    Follicle developed into a wing: *Firmiana colorata* (*Sterculia colorata*)

Fruit 1-seeded with one or more wings
    Aceraceae: *Acer*
    Oleaceae: *Fraxinus*
    Ulmaceae: *Ulmus, Holoptelea*

Fruit a capsule with one sepal foliaceous
    Clusiaceae: *Hymenodictyon excelsum*

Fruit aggregated
  Annonaceae: A bunch of stalked berries
  Magnoliaceae: A bunch of stalked follicles separated on an elongate torus, seeds
    with a red aril

Fruit an utricle
  Verbenaceae: *Tectona*

Spiny Fruits
  Fagaceae: *Castanopsis, Castanea*
  Elaeocarpaceae: *Sloanea*

SEEDS

Seeds winged

Betulaceae	-	*Betula*
Bignoniaceae	-	*Stereospermum, Oroxylum,* etc.
Celastraceae	-	*Lophopetalum*
Coniferae	-	*Pinus, Cedrus, Abies, Picea, Cupressus*
Hamamelidaceae	-	*Altingia*
Hypericaceae	-	*Crotoxylon*
Lythraceae	-	*Lagerstroemia*
Meliaceae	-	*Chukrasia, Toona, Chloroxylon*
Clusiaceae	-	*Neonauclea*
Sterculiaceae	-	*Pterospermum*
Theaceae	-	*Schima*

Nut with a 3-lobed wing

Juglandaceae	-	*Engelhardia*

Seeds arillate

Annonaceae	-	Some
Dilleniaceae		
Euphorbiaceae	-	*Baccaurea**
Clusiaceae	-	Some
Magnoliaceae		
Myristicaceae		
Sapindaceae		

Seeds comose
Apocynaceae - Some
Tamaricaceae - *Tamarix*

---

* *Baccaurea courtallensis* is interesting. Flowers in longish racemes arising from tubercles on
the stem appear as a crimson mass on the whole trunk.

211

Anacardiaceae: Leaves leathery, alternate; juice white, drying black, caustic; stamens often double the number of petals, one in *Mangifera*.

Annonaceae: Flowers trimerous; parts valvate; stamens spirally arranged.

Apocynaceae: Leaves opposite or whorled; milky juice; corolla salver shaped. 5-lobed; stamens five.

Araliaceae: Woody; flowers in an umbel; fruit a berry.

Bombacaceae: Leaves digitate; calyx often subtended by an epicalyx; stamens 5 to numerous, adnate to the base of the corolla and occurring in one or several bundles.

Caesalpinioideae: Leaves pinnate or bipinnate; petals 5, clawed, one somewhat unequal; sepals 5, orbicular, imbricate; fruit a pod.

Cochlospermaceae: Trees with large yellow flowers; anthers 2-locular, dehiscing by terminal pore-like slits; fruit a capsule; seeds covered with woolly hairs.

Asteraceae (Compositae): Rarely shrubs or trees; inflorescence a capitulum supported by overlapping bracts; anthers syngenesious.

Datiscaceae: Tall tree; sepals, petals and stamens four (see *Tetrameles nudiflora*).

Dilleniaceae: Calyx-lobes enlarging in the fruit and becoming fleshy; stamens many; anthers opening by apical pores or lengthwise.

Elaeagnaceae**: Leaves covered on the under-surface with a silvery lepidote indumentum; flowers 4-lobed, lepidote.

Ericaceae: Leaves leathery, alternate; flowers regular, gamopetalous, 5-lobed; stamens double number of lobes; anthers opening by pores.

Clusiaceae: Trees with horizontal branches; leaves opposite, leathery, with fine parallel nerves; juice white or yellow; flowers regular, petals and sepals 2-6; stamens numerous.

Lauraceae: Leaves aromatic; anthers dehiscing by 2 or 4 valves.

**   *Elaeagnus angustifolia* L. var *angustifolia* Nasir, Fl. Pakistan 85:3, 1975 a medium-sized tree is cultivated in Baluchistan for its sweet red globose fruits. It has silvery leaves and fragrant flowers worthy of cultivation in gardens.

Malvaceae: Shrubs, rarely small trees (*Kydia, Thespesia*), commonly herbs; epicalyx characteristic of the family; stamens monadelphous; staminal tube divided at the apex and bearing 1-locular anthers.

Meliaceae: Leaves pinnate; stamens monadelphous (except *Toona*) petals 8-10, valvate; fruit a capsule or baccate.

Mimosoideae: Leaves feathery; flowers minute, almost regular, collected into heads; fruit a pod.

Myrsinaceae: Leaves alternate, punctate; flowers wax-like; sepals and petals often punctate, sometimes black-dotted.

Myrtaceae: Leaves opposite, glandular-punctate; stamens numerous; ovary inferior.

Oleaceae: Stamens 2: leaves opposite; corolla gamopetalous.

Papilionaceae: Flowers sweet pea-like; fruit cupped.

Rhamnaceae: Leaves alternate or opposite; calyx 4-5 lobed; petals 4-5; stamens embraced by the petals.

Clusiaceae: Leaves opposite, with interpetiolar stipules; gamopetalous corolla with inferior ovary.

Sterculiaceae: Stamens monadelphous, divided at apex into filaments and as many staminodes; anthers 2-locular.

To develop an interest in dendrology it is desirable to visit reputed Botanic Gardens, such as the Indian Botanic Gardens, Calcutta; Botanical Garden & Arboreta of the Forest Research Institute, Dehra Dun; National Botanic Research Institute, Lucknow; Lloyd Botanic Garden, Darjeeling; Lal Bagh Botanic Garden, Bangalore; Royal Botanic Garden, Kathmandu; Botanic Garden, Government College, Lahore; Botanic Garden, Peradeniya, Sri Lanka; Botanic Garden, Bogor, Indonesia; and Royal Botanic Gardens, Kew, England.

Several trees in the campus of the Old Rangers College in Dehra Dun where forestry training was initiated are over 100 years old and survive as splendid specimen trees such as the gab (*Diospyros malabarica*) with new leaves turning red, when it looks spectacular; *Podocarpus gracilior* a grand old centenarian, perhaps the oldest in India, is now a massive tree with a rounded crown.

The Arboretum of the Doon School, Dehra Dun, which formerly housed the then Imperial Forestry Research Institute & College has some excellent big trees of immense girth and height. Some of the big trees are the lampati (Devil's tree with

213

girth at breast height of 5.6 m).

In the Museum of the Forest Research Institute, Dehra Dun, there is a 2.5 m wide section of deodar more than 750 years old, dating back to the 13th century when the Qutub Minar was built. Also growing in this Institute is the giant bamboo *Dendrocalamus giganteus* from Myanmar, the world's largest. It towers to 30 m or more and has a 25 cm diameter. During World War II, sections of its culms were used as petrol containers, which were paradropped from Flying Fortresses to camouflaged petrol dumps in the rainforests of Myanmar and beyond, in support of forward elements of the South East Asia Command (SEAC). The growth of this Burmese bamboo is extraordinarily rapid, up to a metre a day being on record.

A walk in the Palmetum of the Calcutta Botanic Garden leads you into an older world of primitive plants – cycads, palms, tree ferns, maiden hair trees – a world of pre-historic plants. These trees constitute the flora of a geological past brought to life in the recent film Jurassic Park, showing the ferocious dinosaurs. The National Botanic Research Institute, Lucknow has *Welwitschia mirabilis,* the strangest of all gymnosperms, with a very short stem. It grows for over a hundred years in the deserts of southwest Africa, where there is a mere trifle of rainfall, the moisture being derived from sea fogs, which cause a heavy deposit of dew. Seeds are produced in profusion. The stem is 2-lobed, which narrows into a tap-root. The leaves are long and ribbon-like. The plants are dioecious and are pollinated by insects (Pearson, *Phil. Trans. R. Soc.* B. 198:291, 1906).

Madan Tamang

Plate 1.a: Campbell's Magnolia *Magnolia campbellii*

K.C. Pradhan

Plate 1.b: White Champa *Michelia doltsopa*

Dev Raj Agarwal

Plate 2: Red Silk Cotton *Bombax ceiba*

M.R. Almeida

Plate 3: Karaya *Sterculia urens*

M.R. Almeida

Plate 4: Bonfire Tree *Firmiana colorata*

Plate 5: Rudraksh *Elaeocarpus augustifolius*

Plate 6: Indian Laburnum *Cassia fistula*

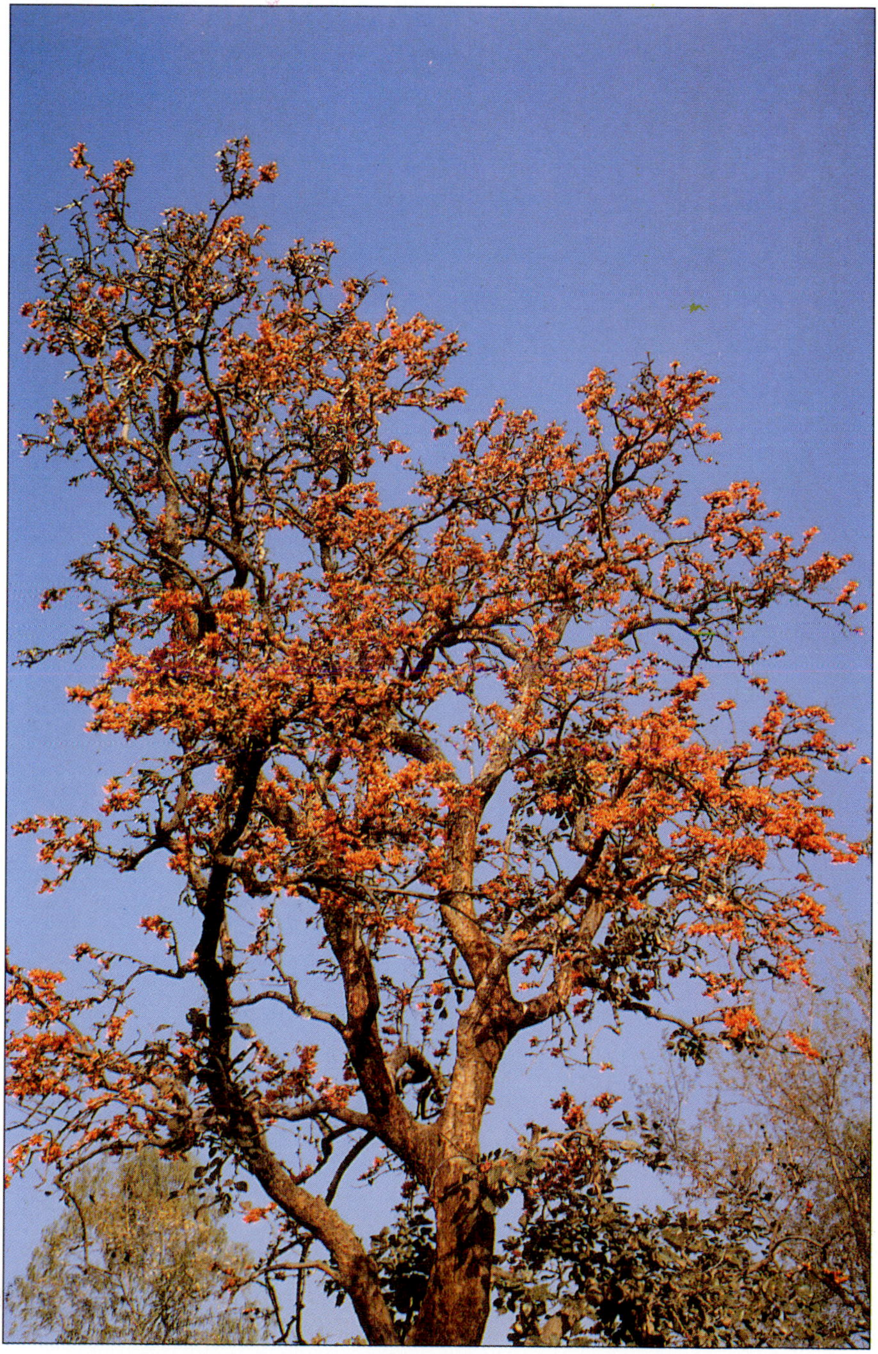

Dev Raj Agarwal

Plate 7: Flame of the Forest *Butea monosperma*

Plate 8: Hollock *Terminalia myriocarpa*

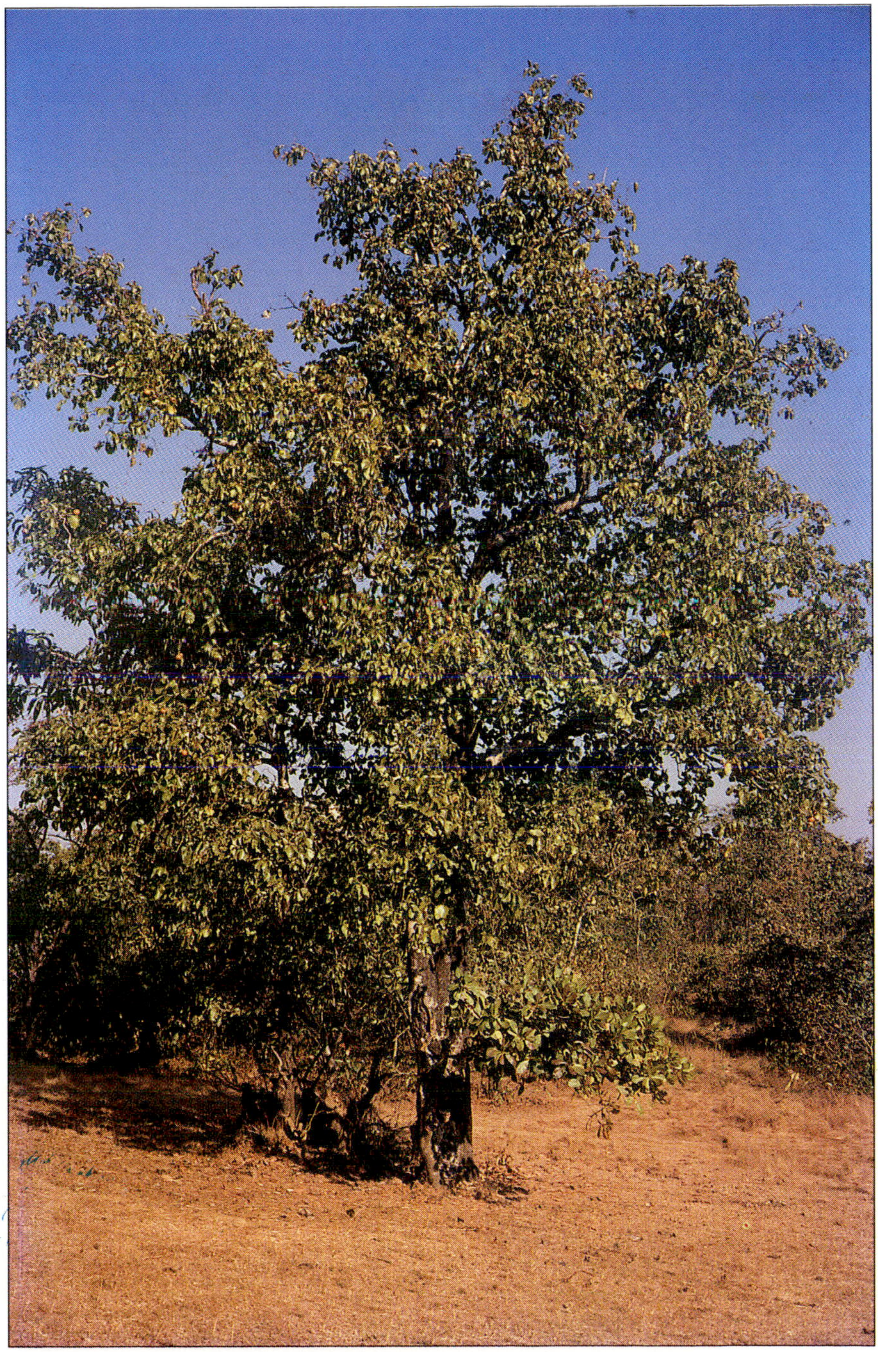

M.R. Almeida

Plate 9: Strychnine Tree *Strychnos nux-vomica*

Plate 10: Tiger Tree *Bischofia javanica*

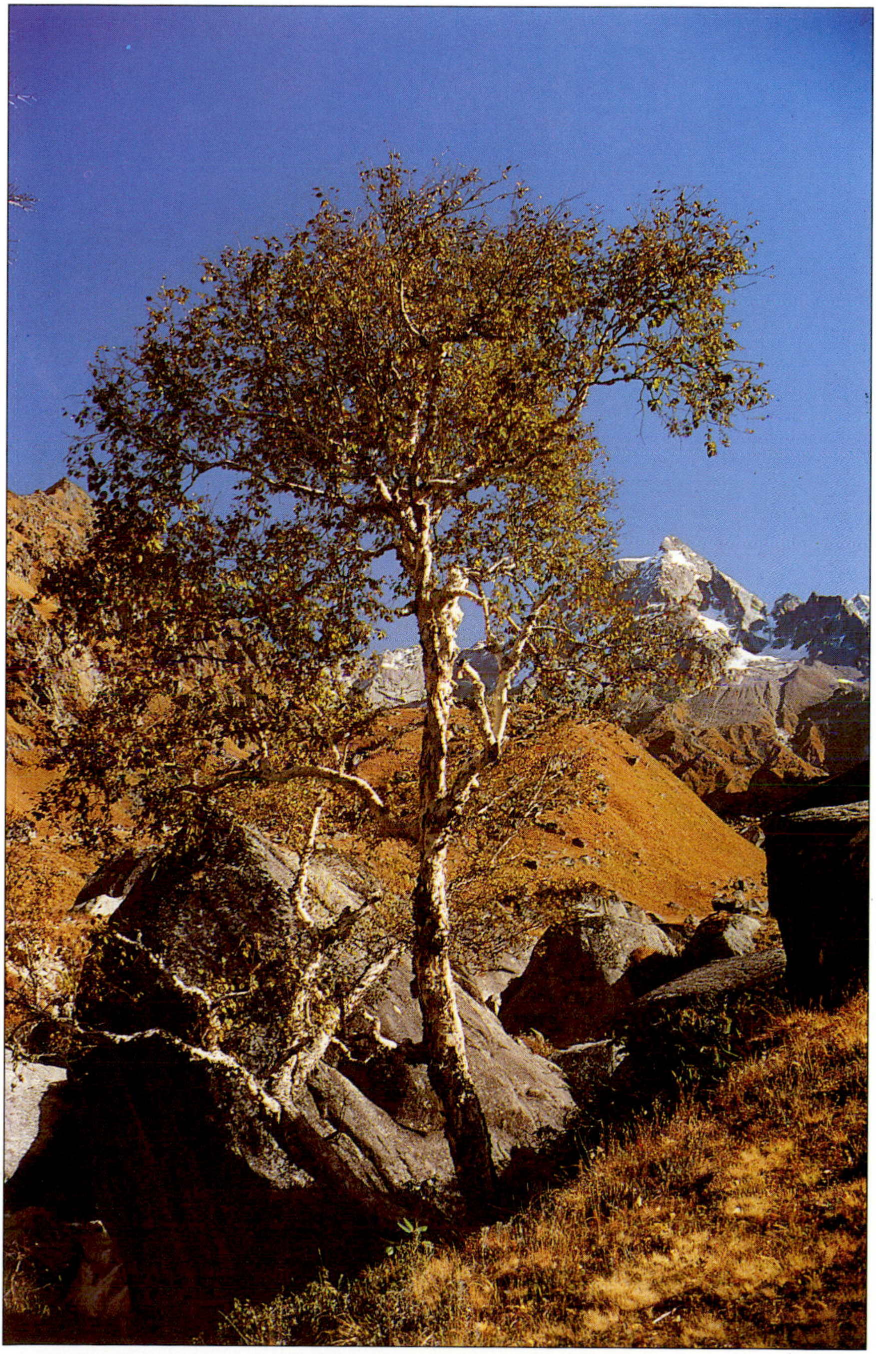

Dev Raj Agarwal

Plate 11: Silver Birch *Betula utilis*

Dev Raj Agarwal

Plate 12: Tiger Bamboo *Bambusa vulgaris*

Dev Raj Agarwal

Plate 13: Chir Pine *Pinus roxburghii*

Plate 14: Chilgoza Pine *Pinus gerardiana*

Plate 15: East Himalayan Fir *Abies densa*

Dev Raj Agarwal

Plate 16: West Himalayan Spruce *Picea smitheana*

a. Date Palm *Phoenix dactylifera*

b. Palmyra Palm *Borassus flabellifer*

c. Fish Tail Palm *Caryota urens*

d. Coconut Palm *Cocos nucifera*

Plate 17. Silhouettes of Palms

a. Fir *Abies pindrow*

b. Himalayan Larch *Larix griffithiana*

c. Drooping Juniper *Juniperus recurva*

d. Black Juniper *Juniperus indica*

Plate 18. Silhouettes of Evergreens

a. Horse Chestnut *Aesculus assamica*

b. Gurjan *Dipterocarpus turbinatus*   c. Red Silk Cotton *Bombax ceiba*

Plate 19. Silhouettes of deciduous trees

Plate 20: *Bruguiera gymnorhiza* buttressed trees with
a few *Rhizophora mucronata* with stilt roots

GLOSSARY

Achene .... dry indehiscent fruit containing only one seed.

Aculeate .... armed with prickles.

Adaxial .... adjacent to the axis.

Adnate .... united to a member of another series.

Androecium .... the male parts of a flower, the stamens as a whole.

Annular .... ring like.

Anther .... the part of a stamen that contains the pollen.

Apiculate .... ending abruptly in a short point.

Apocarpous .... having separate carpels.

Aril .... external growth around a seed, nearly always fleshy, often coloured.

Articulate .... jointed.

Auricle .... ears.

Axillary .... growing at the axil.

Baccate .... berry like.

Berry .... fleshy indehiscent fruit which may contain one or more seeds.

Bifid .... divided in two longitudinally.

Bilabiate .... 2-lipped.

Bipinnate .... when the primary divisions (pinnae) of a pinnate leaf are themselves pinnate.

Bisexual .... with 2 sexes.

Bract .... modified leaf subtending the flower stalk or flower.

Broad leafs .... term used to denote all trees that have a relatively broad leaved blade in contrast to those with needlelike leaves.

Caducous .... falling off.

Calyptra .... petals united into a cup (see Myrtaceae).

Calyx .... outer envelope of a flower of the Dicotyledons, often green or scabrous.

Capitate .... head-like.

Capsule .... dry fruit (composed of more than one carpel) which opens.

Catkin .... a close bracteate, often pendulous, spike.

Caudate .... tailed.

Compound .... formed of several parts; a leaf made of a number of leaflets set on the same axis.

Connate .... when similar parts are united.

Coriaceous .... leathery.

Corolla .... the inner envelope of the flower (petals).

Corymb .... a flat topped inflorescence.

Culm .... stem of grasses and bamboos.

Cuneate .... wedge shaped.

Cuspidate .... tipped with short rigid point.

Cyme .... an inflorescence in which the primary axis bears a single terminal flower that develops first, the system being continued by the axes of secondary and higher order, each with a flower.

Deciduous .... not evergreen, falling off eventually.

Dehiscent .... opening spontaneously.

Deltoid .... triangular.

Dichotomous .... branching regularly into two.

Digitate .... diverging from the same point like the fingers of a hand.

Dioecious .... with male flowers on one individual, female on another.

Distichous .... arranged in two vertical tanks.

Drupe .... a stone fruit.

Emarginate .... notched apex of leaf.

Endemic .... native to particular and usually restricted area.

Entire .... even margin.

Epiphyte .... a plant that grows on another without being parasitic on it.

Fastigiate .... with branches erect more or less appressed.

Flabellate .... fan-shaped.

Follicle .... a several-seeded fruit, resulting from a separate carpel opening by one suture.

Frond .... the leaves of ferns and palm trees.

Gamopetalous ...... a corolla with the petals partially or wholly united.

Gamosepalous ...... sepals more or less united.

Glabrous .... hairless and smooth.

Glandular .... clothed with glands.

Glaucous .... a pale bluish green, or with a pale bloom.

Gynaecium .... the female parts of a flower.

Gynobasic .... a style which arises from near the base of carpels.

Gynophore .... a stalk supporting the ovary.

Heterophyllous ..... having leaves of different shapes on the same plant.

Indehiscent .... not opening.

Induplicate .... folded inwards.

Inflorescence .... arrangement of flowers on the stem.

Integument .... defensive tissues forming an outer covering of various organs.

Internode .... portion of stem between two nodes.

Interpetiolar .... between the petioles.

Intrapetiolar .... within the petiole.

Introrse .... facing inwards.

Kernel .... edible part of a nut (see cashew nut); a seed containing a large amount of reserve nutrients.

Lamina .... leaf blade.

Latex .... milky juice which flows from the tissue of some plants when they are cut.

Lenticels .... any of the raised pores in the stems of woody plants that allow gas exchange between the atmosphere and the internal tissues.

Lepidote .... clothed with scales.

Ligule .... projection from the inside junction of sheath and blade of grasses (including bamboos).

Midrib .... central axis of compound leaf.

Monadelphous ..... in one bundle.

Mucronate .... having a hard sharp point.

Node .... the point at which buds and leaves arise from a stem or branch.

Ovate................... in the shape of an inverted egg, point downward.

Ostiole.................. an opening; specifically, the apical opening in the inflorescence of *Ficus*, or syconium.

Palmate .... divided into 5 segments like the palm of a hand.

Panicle .... a branched raceme.

Pedicel .... the stalk that joins the flower to the point where it is attached to the stem (flower stalk).

Peduncle .... the stalk that joins an inflorescence to the stem.

Pellucid .... translucent.

Pendulous .... hanging down.

Petiole .... the stalk that joins the lamina of a leaf to the stem on which it is growing.

Perianth............... the floral envelope or envelopes (see Apetalae and Monocotyledons).

Pinna .... a primary division of a pinnate leaf.

Plicate .... folded like a fan.

Pollen .... granules formed in the anthers and producing the male gametes.

Pollination............ the transfer of pollen to the stigma from anther.

Polypetalous......... petals free from each other.

Pubescent .... covered with short, soft hairs.

Punctate .... dotted.

Racemose .... inflorescence with stalked flowers on a single axis.

Spathe .... large bract, often coloured or membranous (see palms).

Stamens............... the organs composing the androecium, which carry the anthers containing the pollen, generally at the tip of a filament.

Staminodes .... barren stamens.

Stellate .... star-like.

Stigma.................. the tip of the style on which pollen grains adhere during pollination.

Stipe .... the stalk of a carpel or pistil.

Stipitate .... having a stipe.

Stipule .... a small leaf-like appendage to a leaf usually at the base of a leaf-stem.

Strobiles, cones .... the inflorescence of gymnosperms with unisexual flowers, composed of scales bearing the pollen sacs or of carpellary leaves which carry the ovules, and later, the seeds.

Style .... the narrow part of the pistil bearing the stigma.

Superior .... a condition in which ovary is placed above other floral parts.

Syconium .... inflorescence typical of the genus *Ficus*, formed by a fleshy receptacle, generally pear-shaped, which contains the unisexual flowers and houses the infructescence, composed of little achene-like fruit.

Syncarp .... compound fruit formed by the fusion of the carpels to form a fleshy axis; characteristic of some tropical plants (see *Pandanus*).

Terete .... cylindrical.

Tomentose .... densely covered with short soft hairs.

Torulose .... cylindrical with constrictions or swelling at intervals.

Trifid .... divided into three parts or lobes.

Trifoliate .... three leaved, having three leaflets.

Trimerous .... floral parts in threes.

Tubercles .... excrescences which may be of various shapes or consistency.

Umbel .... a racemose inflorescence in which flower stalks arise from one point.

Umbonate .... (of cone scales) having an umbo, a prominent outgrowth, such as was found in the centre of a warrior's shield.

Undulate .... wavy on the margin (see *Tecomella undulata*).

Unisexual .... of one sex.

Urceolate .... pitcher-shaped, with a rounded body and narrow opening.

Viscid .... sticky.

Zygomorphic .... irregular, flowers which may be divided into equal halves in only one place.

BIBLIOGRAPHY

Readers with access to a botanical library may like to refer to some of the more important works on trees listed below for furthering their interest in trees. For most of the States Forest Flora or Regional Flora are available.

Anonymous (1989). **Biosphere Reserves in India**. Government of India, Ministry of Environment & Forests, New Delhi.

Anonymous (1948-1992). **The Wealth of India - Raw Materials**, 11 Volumes, CSIR, New Delhi.

Bailey, L.H. (1950). **Encyclopedia of Horticulture**. The Macmillan Company, New York.

Bhattacharjee, P.C. *et al.* **Scholar Tree, nesting tree for Adjutant Stork.** Sunday Issue, Times of India 9.1.1994.

Blatter, E. and W.S. Millard (1937). **Some Beautiful Indian Trees** (Revised Edn. by W.T. Stearn, 1977). Bombay Natural History Society, Mumbai.

Bole, P.V. and Y. Vaghani (1986). **Field Guide to Common Indian Trees**. Oxford University Press, Mumbai.

Bor, N.L. (1953). **Manual of Indian Forest Botany**. Oxford University Press, Mumbai.

Brandis, D. (1921). **Indian Trees**. Constable & Co. Ltd., London. (Repr.).

Dassanayake, M.D. and F.R. Fosberg (1980-1985). **A Revised Handbook to The Flora of Ceylon**. V Vols. (incomplete). Oxford & IBH Publishing Co., New Delhi.

Grierson, A.J.C. and Long, D.G. (1983). **Flora of Bhutan** Vol I, Part 1 Royal Botanic Garden Edinburgh.

Hawkins, R.E. (1986). **Encyclopedia of Indian Natural History**. Bombay Natural History Society, Oxford University Press, Mumbai.

Hora, B. (1981). **The Oxford Encyclopedia of Trees of the World**. Oxford University Press, Oxford.

Kanjilal, U.N. *et al.* (1934-1940). **Flora of Assam**. 5 Vols. Shillong. State Floras, District Floras published recently by the Botanical Survey of India, are now available for Howrah, Meghalaya, Jawai, etc.

Kurz, S.A. (1877). **Forest Flora of British Burma**. Vol. I & II. Repr. 1974. B. Singh & M.P. Singh, Dehra Dun.

Macoboy, S. (1986). **Trees for flower and fragrance**. Lansdowne Press, Sydney.

Nayar, M.P. (1985). **Meaning of Indian Flowering Plant Names**. B. Singh & M.P. Singh, Dehra Dun.

Parker, R.N. (1918). **A forest Flora of the Punjab, Hazara and Delhi**. Lahore. (Repr.)

Parker, R.N. (1933). **Common Indian Trees and How to know them**. Manager of Publications, Delhi.

Parkinson, C.E. (1923). **A Forest Flora of the Andaman Islands**. Government Printing Press, Shimla (Repr.)

Sahni, K.C. (1953). **Botanical Exploration of the Great Nicobar Island**. Indian For. 79(1):3-16. Illustrated.

Sahni, K.C. (1990). **Gymnosperms of India and adjacent countries**. Bishen Singh Mahindra Pal Singh, Dehra Dun.

Sahni, K.C., H.B. Naithani, Surendra Singh, S. Biswas, and Babul Das (1996). **Trees of Chandbagh, Doon's National Heritage**. Under the auspices of the Doon School, Dehra Dun. Konark Publishers Pvt. Ltd., Delhi.

Raizada, M.B. and K.C. Sahni (1960). **Living Indian Gymnosperms**. Indian For. Rec. (Botany) 5(2):73-105.

Randhawa, M.S. (1965). **Flowering Trees**. National Book Trust, New Delhi.

Saldanha, C.J. *et al.* (1984). **Flora of Karnataka**. Vol. I. Oxford & IBH Publishing Co., New Delhi.

Stainton, J.D.A. (1972). **Forests of Nepal**. Hafner Publishing Company, New York.

Talbot, W.A. (1909), 1911). **Forest Flora of the Bombay Presidency and Sind**. Vols. I & II. Poona (Repr.).

Troup, R.S. (1921). **Silviculture of Indian Trees**. Vols. I, II & III, Clarendon Press, Oxford. Reprinted.

Venkatesh, C.S. (1976). **Our Tree neighbours**. National Council of Educational Research and Training, New Delhi.

Zhang, J. (Chang King-Wai) (1982). **The Alpine Plants of China**. Science Press, Beijing, China. Gordon and Breach, Science Publishers, Inc., New York.

Index of Scientific Names

A

Abies densa 201, **pl 15**
A. pindrow 200-1, **pl 18**
A. spectabilis 201
Acacia nilotica 76, **77**
Acer caesium 66, **67**
Aceraceae 66
Adansonia digitata 42, **43**
Adina cordifolia 114, **115**
Aegle marmelos 49, **50**
Aesculus assamica 65, **pl 19**
A. punduana 65
Albizia lebbeck 80, **81**
Alnus nepalensis 169, **170**
Alsophila gigantea 206
A. glabra 206
Alstonia scholaris 130, **131**
Amherstia nobilis 82-3
Anacardiaceae 66, 69, 71, 73
Anacardium occidentale 66, **68**
Annonaceae 20
Anthocephalus cadamba 113
A. sinensis 113
Antiaris toxicaria 163
Apocynaceae 130, 132
Aquilaria agallocha 142
A. malaccensis 142, **143**
Areca catechu 184
Arecaceae 180-5
Artocarpus chaplasha 32, 159, **160**
A. heterophyllus 159, **161**
A. integrifolius 159
Azadirachta indica 55, **56**

B

Bambusa arundinacea 186
B. bambos 186
B. spinosa 186
B. vulgaris 188-9
B. vulgaris var. *striata* 189, **pl 12**
B. vulgaris var. *vittata* 188
Bambuseae 186, 188-92
Bauhinia variegata 84
Benthamidia capitata 113
Betula utilis 169, **170, pl 11**
Betulaceae 169
Bignoniaceae 132, 134
Bischofia javanica 148, **149, pl 10**
Bombacaceae 37, 39, 40, 42, 212
Bombax ceiba 39, **pls 2, 19**
B. insigne 39
Borassus flabellifer 182-3, **pl 17**
Boswellia serrata **54**, 55
Bruguiera gymnorhiza 9, **102**, 103, **pl 20**
Burseraceae 53, 55
Butea monosperma 87, **pl 7**

C

Caesalpinioideae 80
Calophyllum elatum 28
C. polyanthum 28
C. tomentosum 28
Canarium strictum 53
C. resiniferum 53
Capparaceae 23
Caryota urens 183, **pl 17**
Cassia fistula 84, **pl 6**
Casuarina equisetifolia 167, **168**
Casuarinaceae 167
Cedrela toona 57
Cedrus deodara 203
Ceiba pentandra 40
Cinnamomum tamala **139**, 140
C. verum 140, **141**, 142
C. zeylanicum 142
Clusiaceae 25-9, 113-4
Cochlospermaceae 22

Cochlospermum religiosum 22
Cocos nucifera 185, **pl 17**
Combretaceae 104, 106-7
Cornaceae 113
Cornus capitata 113
Corypha umbraculifera 181
Crateva magna 23, **24**
C. nurvala 23
C. religiosa 23, 25
Cullenia exarillata 37, **38**
Cupressaceae 204-5
Cupressus torulosa 205
Cyathea brunoniana 205
C. gigantea 206
C. glabra 20
Cyatheaceae 205-6
Cycadaceae 192
Cycas circinnalis 192, **193**

D

Dalbergia latifolia 93, **95**
D. sissoo 96, **97**
Dendrocalamus giganteus 191
D. hamiltonii 190
D. strictus 189
Dillenia indica 16
Dilleniaceae 16
Diospyros ebenum 124, **125**
D. kurzii 126
D. marmorata 126, **127**
D. melanoxylon 122, **123**
Dipterocarpaceae 32, 34-6
Dipterocarpus turbinatus 32, **33**, **pl 39**
Durio zibethinus 40, **41**

E

Ebenaceae 122, 124, 126
Echinocarpus assamicus 47
Elaeocarpaceae 47
Elaeocarpus augustifolius 47, **pl 5**
E. ganitrus 47

E. sphaericus 47
Ēmbilica officinalis 146, **147**
Engelhardia spicata 166
Ericaceae 116
Erythrina indica 88
Erythrina variegata 88
Euphorbiaceae 146, 148, 150
Excoecaria agallocha **148**, 149

F

Fagaceae 171, 173-4
Feronia elephantam 51
Ficus benghalensis **153**, 154
F. elastica **157**, 158
F. religiosa 155, **156**
Firmiana colorata 46, **pl 4**

G

Garcinia indica 25
G. mangostana 26
Gmelina arborea 135, **136**

H

Hopea ponga 36
H. wightiana 36
Hyphaene dichotoma 182

J

Juglandaceae 166
Juglans regia 166
Juniperus indica **pl 18**
J. macropoda 204
J. polycarpos 204
J. recurva **pl 18**

L

Lagerstroemia flos-reginae 111
L. reginae 111
Lannea coromandelica 69, **70**
Larix griffithiana 203, **pl 18**

Lauraceae 140
Limonia elephantum 51, **52**
Lythraceae 111

M

Madhuca longifolia var *latifolia* 117, **118**
Magnolia campbellii 17, **pl 1a**
M. pterocarpa 17
Magnoliaceae 17-9
Malvaceae 36
Mangifera indica 71-2
M. sylvatica 71
Manilkara hexandra **121**, 122
Meliaceae 55, 57
Melocanna baccifera 191
M. bambusoides 191
Mesua ferrea 27, 32, 82
Michelia champaca 18
M. doltsopa 19, **pl 1b**
M. excelsa 19
Milletia ovalifolia 100
Millingtonia hortensis 132, **133**
Mimosoideae 76, 78, 80
Mimusops elengi 119, **120**
M. hexandra 122
Moraceae 154-5, 158-9, 162-3
Moringa oleifera 73, **75**
Moringaceae 73
Morus serrata 162
Myrtaceae 109

N

Nyctanthes arbor-tristis 128

O

Olea europaea 130
O. ferruginea 128, **130**
Oleaceae 128
Ougeinia oojeinensis 98
O. dalbergioides 98

P

Pandanaceae 186
Pandanus odoratissimus 186, **187**
Phoenix dactylifera 180, **pl 17**
P. sylvestris 180
Phyllanthus emblica 146
Picea smithiana 202, **pl 16**
P. spinulosa 202
Pinaceae 195-203
Pinus bhutanica 199
P. excelsa 198
P. gerardiana 196, **pl 14**
P. insularis 197
P. kesiya 197
P. longifolia 195
P. merkusii 199
P. roxburghii 195, **pl 13**
P. wallichiana 198-9
Platanaceae 164
Platanus orientalis 164, **165**
Podocarpaceae 195
Podocarpus latifolia 195
P. wallichianus 195
Poeciloneuron indicum 29, **30**
Polyalthia longifolia 20, **21**
Pongamia pinnata 98, **99**
Populus ciliata 174, **175**
P. euphratica 176, **177**
Prosopis cineraria 78, **79**
P. spicigera 78
Pterocarpus dalbergioides 91, **92**
P. marsupium 89, **90**
P. santalinus 93, **94**
Putranjiva roxburghii 150
Pyrus pashia 101

Q

Quercus incana 171
Q. lamellosa 19, 174
Q. leucotrichophora 171, **172**
Q. semecarpifolia 173, 20

R

Rhamnaceae 59
Rhizophora mucronata 101, **102, pl 20**
Rhizophoraceae 101, 103
Rhododendron arboreum 116
Rosaceae 101
Rubiaceae 113-4
Rutaceae 49, 51

S

Salicaceae 174, 176, 178
Salix alba 178
S. babylonica 178
S. tetrasperma 178, **179**
Santalaceae 144
Santalum album 144, **145**
Sapindaceae 61, 63
Sapindus emarginatus 61, **62**
S. mukorossi 61
Sapotaceae 117, 119, 122
Saraca asoca 22, 80
Schima wallichii 31
Schinus limonia 51
Schleichera oleosa 63, **64**
S. trijuga 63
Semecarpus anacardium 73, **74**
Shorea robusta 32, 34
Sloanea assamica 47, **48**
Sterculia urens 44, **45, pl 3**
Sterculiaceae 44, 46
Strychnos nux-vomica 132, 144, **pl 9**
Syzygium cumini 109, **110**

T

Talauma hodgsonii 19
Tamarindus indica 85, **86**
Taxaceae 194
Taxus baccata 194
Tectona grandis 137, **138**
Tecomella undulata 134
Terminalia arjuna 104, **105**

T. bialata 107
T. chebula 107, **108**
T. crenulata 106
T. myriocarpa 104, **pl 8**
Tetramelaceae 112
Tetrameles nudiflora 112
Theaceae 31
Thespesia populnea 36
Toona ciliata 57, **58**
Tsuga brunonia 202
T. dumosa 201

U

Ulmaceae 151
Ulmus wallichiana 151, **152**

V

Vateria indica 35
Verbenaceae 135, 137

Z

Ziziphus mauritiana 59, **60**

Index of Common Names

A

Alder 4, 5, 11, 169
Amaltas 2, 84
Amla 4, 146
Andaman Marble Wood 126
Andaman Redwood 91
Areca Palm 184
Arjun 104
Arunachal Pine 199
Ashoka 22, 80, 82-3, 155
Assam Rubber Tree 10, 158

B

Babul 2, 6, 9, 76, 210
Bael 2, 49
Ballagi 28-9
Ban Oak 171
Banyan 154-5
Beedi-leaf Tree 122
Betel Nut Palm 184
Bishop Wood 148
Black Dammar 7
Black Plum 109
Blue Pine 4-6, 198-9, 202
Bo Tree 155
Bodhi Tree 155
Bonfire Tree 9, 46
Brown Oak 4, 173
Buk Oak 5, 19, 174, 210
Bullet Wood 11, 119
Butter Tree (Kokam) 25
Butter Tree (Mahua) 117, 119
Buttercup Tree 9, 22

C

Camphor 2, 10
Cashew nut 8, 66, 83

Cassowary Tree 167
Champak 18
Chaplash 10-11, 159
Chilgoza Pine 4, 196
Chinar 14, 164, 208
Chir Pine 4, 195-6, 199
Cinnamon 6, 13, 27, 140, 142
Civet Fruit 40
Coconut Palm 185
Coloured Sterculia 46
Coral Jasmine 128
Coral Tree 7, 11, 88-9
Coromandel Ebony 122
Crepe Myrtle 111
Crozier Cycas 192
Cypress 1, 4, 205, 207

D

Date Palm 154-5, 180, 207
Deodar 4, 5, 203, 214
Durian 13, 40, 42

E

East Himalayan Horse Chestnut 65
Ebony 122, 124
Edible Date Palm 180
Elephant Apple 5, 10, 16, 51
Elm 4, 151
Emblic Myrobalan 146
Euphrates Poplar 178

F

Fan Palm 181-3
Fish Tail Palm 1, 183
Flame Amherstia 14, 82
Flame of the Forest 3, 9, 87

G

Gamari 5, 10, 135, 208
Garlic Pear 23
Giant Bamboo 191, 214
Giant Thorny Bamboo 8, 186
Golden Apple 49
Golden Bamboo 1, 188
Golden Champ 18
Golden Shower 84
Green Striped Bamboo 188
Grey Oak 4, 171
Gurjan 11-2, 32

H

Haldina 4, 8-9, 114
Heart Flower 5, 10, 19
High Level Fir 4, 201
Himalayan Cypress 4, 205
Himalayan Hemlock 6, 201
Himalayan Larch 5-6, 203
Himalayan Long-Needle Pine 195
Himalayan Mulberry 162
Himalayan Pencil Juniper 4, 204
Himalayan Poplar 4, 174
Himalayan Silver Birch 6, 169
Himalayan Walnut 166
Hollock 5, 11, 104
Horse Chestnut 4, 63, 65
Horse Tail Tree 11, 167

I

Incense Tree 53
India Rubber Tree 158
Indian Copal Tree 7, 34
Indian Coral Tree 11, 88
Indian Cork Tree 132
Indian Doum Plant 182
Indian Fan Palm 182
Indian Frankincense Tree 53
Indian Hemlock Fir 201-2
Indian Jujube 6, 59, 207
Indian Kino Tree 89

Indian Laburnum 4, 84
Indian Medlar Tree 119
Indian Olive 128
Indian Poplar 176
Indian Tulip Tree 36
Indian Willow 178
Indus Poplar 6
Ironwood 7, 10

J

Jack Tree 159
Jamun 2, 4, 10, 109
Jarul 111
Jewels on a String 100
Jhingan 69
Jungli dungy 5, 7, 14, 112

K

Kadam Tree 113
Kala dammar 10
Karaya 9, 44, 46
Keora 186
Kharsu Oak 173
Khasi Pine 11, 13, 197
Khejri 78
Kokam 25-6

L

Lampati 213
Large-flowered Dillenia 16
Laurel 8, 11, 106

M

Magnolia 5, 6, 11, 13, 15, 17
Mahua 9, 117
Mango 2, 71, 82, 144, 207
Mangosteen 13, 26
Mangrove 7-9, 13, 101, 103, 148
Margosa Tree 55
Marking-Nut Tree 71
Mast Tree 20, 22, 207
Merkus Pine 199
Monkey Bread Tree 42, 44

Moulmein Rosewood 100
Muli Bamboo 191
Myrobalan 8, 107, 146

N

Needle Wood 29
Neem 55
Nux-vomica Tree 10, 132

O

Oriental Plane 164

P

Padauk 12, 91
Palmyra Palm 182
Piney Tallow 35
Pipal 2, 36, 155, 174
Poison Nut Tree 132
Pongam Oil Tree 98
Poon 28
Poonspar Tree 28
Portia Tree 36
Pride of Burma 82
Pride of India 111
Putranjiva 150-1

Q

Queen's Flower 111

R

Red Sanders 10, 93, 144
Red Silk Cotton 2, 3, 5-6, 8-9, 11-2, 39, 207
Rhino Bamboo 190
Rose Wood 93
Rudraksh 47

S

Sack Tree 163
Sacred Barna 23

Sal 1, 3, 5-6, 8-10, 31-2, 34, 100
Salai 6, 9, 53, 55
Sandalwood 10, 93, 144
Sandan 4, 98, 208-9
Scholar Tree 130
Screw Pine 5, 8, 11, 186-7, 207
Silver-Grey Wood 107
Sissoo 3, 96-7
Snake Wood 132
South Indian Ritha 61
South Indian Soapnut 61
Spar Tree 7, 28
Strawberry Tree 113

T

Talipot Palm 7, 13, 181
Tamarind 2, 85, 87, 207
Teak 2, 6, 9, 13-4, 32, 34, 91, 96, 111, 135, 137, 210
Tenasserim Pine 199
Tiger Tree 5, 148
Toon 4, 57
Torchwood Tree 22
Tree Ferns 1, 5, 9, 12, 214
True Cinnamon Tree 140
True Kapok 39, 40
True Mangrove 101

U

Umbrella Tree 36
Upas Tree 14, 163, 209
Utrasum Bead Tree 47

V

Variegated Bauhinia 84

W

Walnut 166
Wavy-leaved Tecomella 134
West Himalayan Elm 151
West Himalayan Fir 200

West Himalayan Horse Chestnut 63
West Himalayan Spruce 202
Whistling Pine 11, 167
White Chuglam 12, 107
White Dhup 35, 53, 208
White Silk Cotton 39
Wodier Wood 69

Y

Yellow Bamboo 188
Yellow Silk Cotton Tree 9, 22
Yew 1, 4-5, 194

NOTES

NOTES

NOTES

NOTES

NOTES

NOTES

ABOUT THE BOMBAY NATURAL HISTORY SOCIETY

The BNHS was founded in 1883 and today it is the prime non-governmental conservation organisation in the Subcontinent. We work towards the conservation of nature and natural resources, education and research in natural history, and have members in over 25 countries.

Membership Activities and Benefits

- Nature camps to wildlife places both in and outside India.
- Treks, walks and field trips at weekends.
- Excellent audio-visuals presented by experts regularly.
- Seminars, workshops and correspondence courses on specific natural history subjects.
- Members receive *Hornbill*, a quarterly magazine.
- Subscription to the *Journal* is optional to members.
- 25% discount on all BNHS publications.
- 5% discount on BNHS cards and calendars.
- Access to the finest collection of books on natural history.
- Voluntary Nature Education and Conservation activities.

Publications

BNHS Publications have been the standard reference works on the natural history of the Indian subcontinent since 1886. They are essential acquisitions for naturalists, amateurs and professionals throughout the country and abroad. Published uninterrupted since 1886, the *Journal of the Bombay Natural History Society* is acknowledged to be one of the finest scientific natural history sources for the Oriental Region. The popular quarterly magazine *Hornbill*, published since 1976, caters to a varied readership of all ages.

For details contact:

Bombay Natural History Society

Hornbill House, S.B. Singh Road, Mumbai 400 023, Maharashtra, India.
Tel.: +91-22-2282 1811 Fax: +91-22-2283 7615
E-mail: bnhs@bom4.vsnl.net.in Website: www.bnhs.org

THE SOCIETY'S PUBLICATIONS

		List Price
1.	**The Book of Indian Birds** by Sálim Ali, 13th edition	Rs. 495
2.	**A Pictorial Guide to the Birds of the Indian Subcontinent** by Sálim Ali & S. Dillon Ripley, 2nd edition	Rs. 370
3.	**A Guide to the Cranes of India** by Prakash Gole	Rs. 75
4.	**Birds of Wetlands and Grasslands** by Asad R. Rahmani & Gayatri Ugra	Rs. 500
5.	**Birds of Western Ghats, Kokan and Malabar** by Satish Pande, Saleel Tambe, Clement Francis M. & Niranjan Sant	Rs. 950
6.	**Petronia** by J.C. Daniel and Gayatri Ugra	Rs. 400
7.	**The Book of Indian Animals** by S.H. Prater, 3rd edition	Rs. 275
8.	**A Week with Elephants — Proceedings of the Seminar on Asian Elephants, June 1993** Edited by J.C. Daniel & Hemant Datye	Rs. 450
9.	**The Book of Indian Reptiles and Amphibians** by J.C. Daniel	Rs. 595
10.	**The Book of Indian Shells** by Deepak Apte	Rs. 295
11.	**The Book of Indian Trees** by K.C. Sahni, 2nd edition	Rs. 275
12.	**Some Beautiful Indian Trees** by E. Blatter & W.S. Millard	Rs. 295
13.	**Some Beautiful Indian Climbers and Shrubs** by N.L. Bor & M.B. Raizada, 2nd edition	Rs. 525
14.	**Common Indian Wildflowers** by Isaac Kehimkar	Rs. 375
15.	**Illustrated Flora of Keoladeo National Park, Bharatpur** by V.P. Prasad, Daniel Mason, Joy E. Marburger & C.R. Ajithkumar	Rs. 695
16.	**Sálim Ali's India** Edited by A.S. Kothari & B.F. Chhapgar	Rs. 1200
17.	**A Century of Natural History** Edited by J.C. Daniel	Rs. 210
18.	**Encyclopedia of Indian Natural History** Edited by R.E. Hawkins	Rs. 1250
19.	**Conservation in Developing Countries —Problems and Prospects** Edited by J.C. Daniel & J.S. Serrao	Rs. 400
20.	**Cassandra of Conservation** Edited by J.C. Daniel	Rs. 200
21.	**Calls of Indian Birds** set of two audio cassettes with explanatory booklet. Digitally mastered	Rs. 160

FORTHCOMING TITLE

Book of Indian Butterflies
by Isaac Kehimkar

25 % DISCOUNT ON ALL BNHS PUBLICATIONS FOR MEMBERS